# Homocystinuria
## A risk factor of premature vascular disease

## CLINICAL RESEARCH SERIES

The aim of this series is to draw attention to results of clinical research projects of interest to physicians in the field of internal medicine and related disciplines.
The editors hope that by reaching a wide audience, it will increase and facilitate practical application of recent findings.

# Homocystinuria
## A risk factor of premature vascular disease

*G.H.J. Boers*

1986
FORIS PUBLICATIONS
Dordrecht-Holland/Riverton-U.S.A.

*Published by:*
Foris Publications Holland
P.O. Box 509
3300 AM Dordrecht, The Netherlands

*Sole distributor for the U.S.A. and Canada:*
Foris Publications U.S.A.
Riverton N.J. 08077
U.S.A.

ISBN 90 6765 146 X



Printed in the Netherlands by ICG Printing, Dordrecht.

# Contents

# Preamble

In 1977 a 21-year old woman suffering from ectopia lentis had been referred to our clinic to exclude the presence of a systemic disorder which might have caused her ocular abnormality. At physical examination ischemia of the toes of the left foot and an abdominal bruit were established. Angiographic examination revealed multiple and severe, partial or complete stenoses in abdominal arteries. Biochemical and enzymatic investigations established the diagnosis of homozygous homocystinuria. Furthermore, screening of her sibs for the presence of homocystinuria led to the diagnosis in one of her sisters at the age of 29, in whom severe stenoses of both carotid arteries had been established elsewhere already at the age of 26 when she suffered from severe psychiatric illness leading to attempted suicide. After the diagnosis in these 2 sisters, homocystinuria had been established in another 5 patients until 1979 and the data in these first 7 patients were reported in 1980[1]. Up to the end of 1984 the number of homozygous homocystinuria patients has been extended to 21. Their biochemical, enzymatic, pathological, and therapeutic data are the subject of the first part of this thesis.

The observation of striking vascular lesions in these homozygous patients leading to severe thromboembolic events at very young age, challenged us to explore the possibility whether even mild homocysteinemia, as present in carriers for this condition, might be a risk factor of premature cardiovascular disease. This would indicate a broader role of such heterozygosity in the overall pathogenesis of arteriosclerosis in the general population in view of the estimated frequency as high as 1 in 70 to 1 in 200 of heterozygotes. A prerequisite for the detection of homocystinuria heterozygosity in patients with premature arteriosclerosis was the availability of a valid method to identify such heterozygotes. This led to a re-evaluation of the sensitivity of the earlier described procedure of a standardized oral methionine loading test[2] to discriminate heterozygotes from subjects with a normal status. Results from such tests performed in a large number of normal control subjects disclosed statistically significant differences in methionine handling between pre- and postmenopausal women and men. Under the proviso of taking these differences into account, it proved possible to discriminate obligate carriers for homocystinuria from normal controls in 90% by means of methionine loading. By the use of a similar test heterozygosity for homocystinuria then could be established in about 30% of patients suffering from premature arteriosclerosis which had caused occlusive peripheral or cerebral arterial disease. In these patients heterozygosity was confirmed by determination of cystathionine synthase activity in cultured skin fibroblasts. Remarkably, in patients suffering from myocardial infarction at young age, the frequency of heterozygosity was found no higher than in the general population. These observations led to the final conclusion in part II of this thesis that even heterozygosity for homocystinuria due to cystathionine synthase deficiency must be considered as a risk factor of premature arteriosclerosis especially of peripheral and cerebral vessels.

REFERENCES

1. Drayer JIM, Cleophas AJM, Trijbels JMF, Smals AGH, Kloppenborg PWC. Symptoms, diagnostic pitfalls, and treatment of homocystinuria in seven adult patients. Neth J Med 1980; 23: 89–94.
2. Sardharwalla IB, Fowler B, Robins AJ, Komrower GM. Detection of heterozygotes for homocystinuria. Arch Dis Child 1974; 49: 553–9.

# Part I
# Homozygous homocystinuria

*Chapter 1*

# State of the art

## 1.1 GENERAL INTRODUCTION

In 1962 Carson and coworkers, during screening tests of mentally retarded patients in Northern Ireland, discovered two sisters to be excreting homocystine, a sulphur containing amino acid not detected in normal urine[1,2]. Independently, Gerritsen reported at the same time data of a male mentally retarded patient with homocystinuria in Wisconsin[3]. In a short time thereafter, the number of studied homocystinuria patients had been sufficiently extended to delineate the major biochemical and clinical characteristics of homocystinuria[4,5]. In 1964 Mudd demonstrated in a typical patient a specific deficiency of hepatic cystathionine synthase, a pyridoxine dependent enzyme on the transsulphuration pathway which promotes the conversion of the essential amino acid methionine into cysteine[6]. Thus, homocystinuria had been clinically outlined and was established as an inborn error of metabolism within 2 years. Up to the present more than 600 patients with homocystinuria due to cystathionine synthase deficiency have been diagnosed[7].

Since 1969 it was recognized that the phenotype "homocystinuria" can be caused by different genotypes[8]. Apart from cystathionine synthase deficiency at least two other enzyme defects result in homocystinuria, each involved in the re-methylation pathway from homocysteine back to methionine. These conditions are extremely rare and no more than about 35 patients have been reported so far[9-13]. Therefore, albeit not appropriate to use the term "homocystinuria" without the definition of the genotype, it is common usage to regard it as synonymous with "cystathionine synthase deficiency".

## 1.2 METHIONINE METABOLISM

The reactions of the transsulphuration pathway converting methionine to cysteine and inorganic sulphate and the related metabolic reactions are shown in Figure 1.

The first step in the metabolism of methionine is the conversion to S-adenosyl-methionine by methionine adenosyltransferase. S-adenosyl-methionine is the major methyl donor in mammals. All transmethylations by S-adenosyl-methionine result in the formation of S-adenosyl-homocysteine, which in turn is hydrolyzed to homocysteine and adenosine by S-adenosyl-homocysteine hydrolase. Homocysteine lies at a metabolic branch point from which it may be transsulphurated to cystathionine or re-methylated to reform methionine. On normal diet the available homocysteine in man is nearly equally divided between the transsulphuration and re-methylation pathway. However, variations in methionine intake are balanced by changing this equilibrium[14].

The transsulphuration of homocysteine starts with the condensation of homocysteine with serine to form cystathionine. This reaction is catalyzed by cystathionine synthase. This enzyme requires pyridoxal phosphate as cofactor. Under in vivo conditions the equilibrium of this reaction is very much toward cystathionine formation. By high protein intake the activity of cystathionine synthase is increased[15,16]. Cystathionine synthase deficiency is the most frequent cause of the homocystinuria syndrome, as mentioned earlier. Cystathionine is cleaved into cysteine and homoserine by γ-cystathionase, another pyridoxine dependent enzyme. Various enzymes metabolize cysteine further to the major products of sulphur metabolism, taurine and sulphite which is oxidized by sulphite oxidase to sulphate.

The re-methylation of homocysteine back to methionine in man is enabled by two alternative reactions. In one, betaine – a choline derivative – donates the methyl group to homocysteine in a reaction catalyzed by betaine-homocysteine methyltransferase. Animal experiments showed an increase of the activity of this enzyme in the case of limited as well as excessive methionine intake[17], and an increase of the activity by diets supplemented with choline or betaine[18]. The other homocysteine re-methylating enzyme is $N^5$-methyltetrahydrofolate-homocysteine methyltransferase which in the case of normal protein intake accounts for about 50% of the re-methylation of homocysteine[16]. High methionine intake decreases its activity[19]. The cosubstrate in this reaction, 5-methyltetrahydrofolate, is formed by another enzyme, 5,10-methylenetetrahydrofolate reductase. About 20 patients have been reported up to now with homocystinuria due to deficient activity of this latter enzyme, leading to a blockade of the re-methylation of homocysteine due to lack of the cosubstrate[9]. The metabolism of vitamin $B_{12}$ interacts with the metabolism of methionine and folate as shown in Figure 1 where it is illustrated that methylcobalamin acts as the cofactor of $N^5$-methyltetrahydrofolate-homocysteine methyltransferase which re-methylates homocysteine to methionine. Furthermore, vitamin $B_{12}$ is cofactor in the enzymatic conversion of methyl-

# THE METABOLISM OF METHIONINE IN MAN

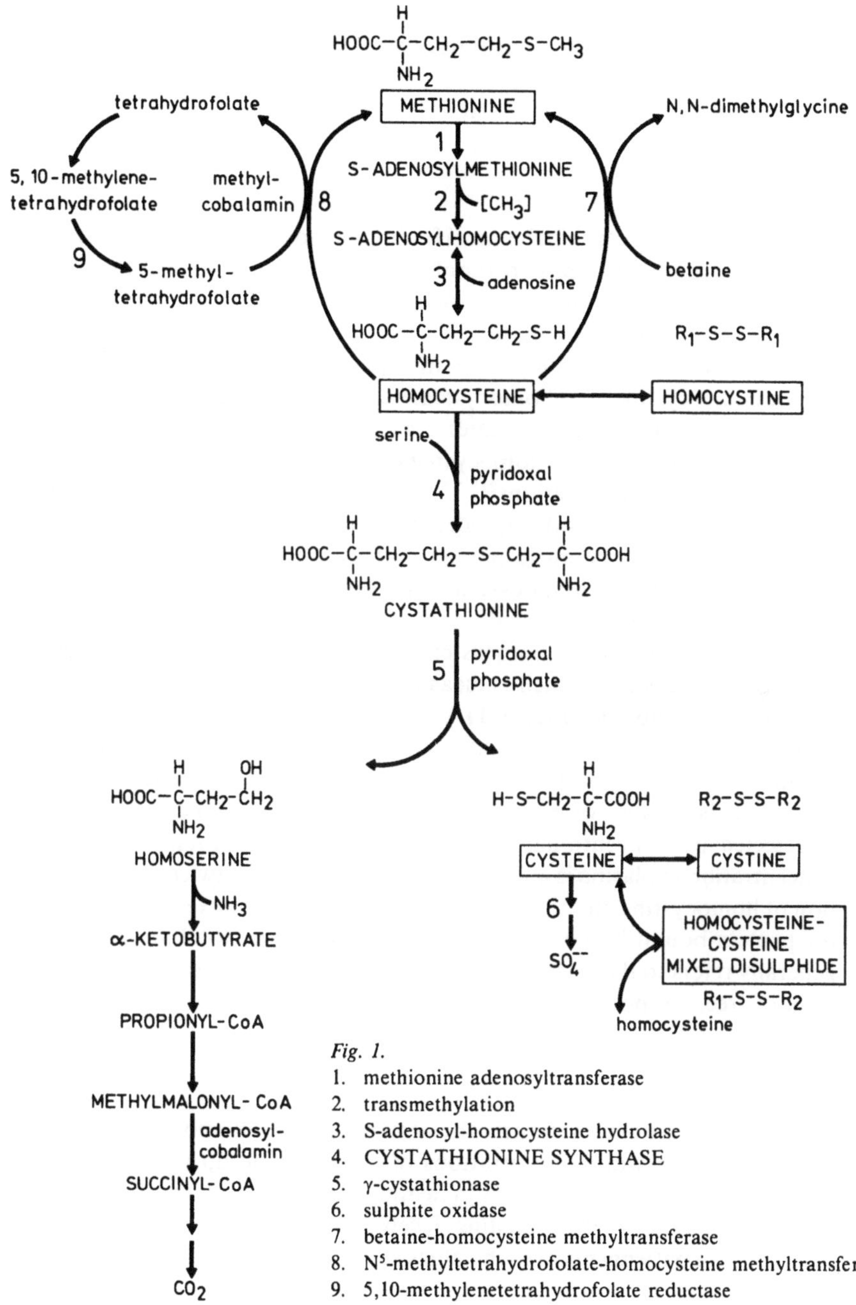


*Fig. 1.*

1. methionine adenosyltransferase
2. transmethylation
3. S-adenosyl-homocysteine hydrolase
4. CYSTATHIONINE SYNTHASE
5. γ-cystathionase
6. sulphite oxidase
7. betaine-homocysteine methyltransferase
8. $N^5$-methyltetrahydrofolate-homocysteine methyltransferase
9. 5,10-methylenetetrahydrofolate reductase

malonyl-CoA to succinyl-CoA. Up to the present 15 patients with homocystinuria due to defects in the formation of these cobalamin cofactors have been detected resulting in deficient re-methylation of homocysteine[10-13]. In all but one patient[12] methylmalonic aciduria was established in addition.

## 1.3 BIOCHEMICAL CHARACTERISTICS

### *1.3.1 General remarks*

The most characteristic biochemical findings in homocystinuria due to cystathionine synthase deficiency are the abnormal accumulation of homocysteine and methionine in combination with decreased conversion of methionine to cysteine and inorganic sulphate.

Homocysteine and cysteine are in equilibrium with their disulphides, homocystine, cystine, and homocysteine-cysteine mixed disulphide. Intracellularly homocysteine predominates over its disulphide because of a disulphide interchange between homocysteine and reduced glutathione[20] which is present in high concentrations[21]. Using special techniques the presence of homocysteine besides its disulphides has been demonstrated in fresh blood and urine of homocystinuria patients[22,23]. Nevertheless with conventional procedures to measure homocysteine, it is rapidly oxidized during blood and urine sampling and amino acid analysis and is measured only as its disulphides, homocystine and homocysteine-cysteine mixed disulphide. This also holds true for cysteine which is measured as cystine and homocysteine-cysteine mixed disulphide (Figure 1).

It has been shown recently that a portion of homocysteine and cysteine in plasma is bound to proteins which can be released by 2-mercapto-ethanol[24,25]. This amount of protein-bound homocysteine and cysteine remains undetected by conventional amino acid analysis. The total homocysteine concentration in plasma is in fact recovered as the sum of twice the concentration of homocystine and that of homocysteine-cysteine mixed disulphide and the protein-bound homocysteine. However, the term "total homocysteine" generally refers to the former two concentrations and represents virtually the total of free, i.e. not protein bound homocysteine.

### *1.3.2 Blood*

In patients with homocystinuria the accumulation of homocysteine leads to homocystinemia with concentrations of homocystine up to 200 µmol/l in fasting plasma or serum whereas in fasting normal subjects free homocystine is not detectable[22,23,26,27]. Fasting levels of homocysteine-cysteine mixed disulphide in patients may reach values as high as 55 µmol/l[27], whereas these concentrations in healthy persons are very low ranging from 0.2 to 5.6

µmol/l[28,29]. Fasting methionine levels in homocystinuria due to cystathionine synthase deficiency may range from normal, i.e. about 30 µmol/l, to as high as 2000 µmol/l[22,23,26,27]. The hypermethioninemia, present in about 94% of the patients[7], results from enhanced re-methylation of homocysteine rather than from underutilization of methionine.
Cystine, a substance distal to the metabolic block located at cystathionine synthase, is formed at an abnormally slow rate in homocystinuria due to deficiency of this enzyme. In these patients the plasma or serum levels are low-normal, reduced or even undetectable[27,30-32]. However, the abnormally low cystine levels are partially due to the formation of homocysteine-cysteine mixed disulphide in the presence of accumulated homocysteine[27,33].

In contrast to homocysteinemia due to cystathionine synthase deficiency, the homocysteinemia due to 5,10-methylenetetrahydrofolate reductase deficiency or impaired synthesis of methylcobalamin is not accompanied by hypermethioninemia and lowered cystine levels. The methionine levels in these patients were normal or even decreased with cystine levels in the normal range or in lowered concentrations[9-13].

### *1.3.3 Urine*

The tubular reabsorption of homocystine is rather low as compared with most other amino acids, ranging in studied homocystinuria patients from 80% to 97%[26,32], and therefore the homocystinemia in these patients is virtually always accompanied by homocystinuria with urinary homocystine excretions up to 1000 µmol per day[22,26] whereas in urine of normal subjects homocystine is not detected. Homocysteine-cysteine mixed disulphide in urine of homocystinuria patients is measured in concentrations of 5%-10% of the corresponding homocystine levels[23]. In addition, more than 20 other metabolites of homocysteine have been identified in the urine of homocystinurics[34].
Hypermethioninemia in patients with homocystinuria due to cystathionine synthase deficiency does not obligatorily lead to an increase of urinary methionine excretion because of nearly complete renal tubular reabsorption[26]. Urinary methionine content within the normal range has been reported in cystathionine synthase deficiency with even moderate elevations of the plasma methionine level[4,26].
The urinary excretion of cystathionine and cystine in patients with homocystinuria due to cystathionine synthase deficiency is lowered[35,38], and low or normal taurine excretion has been reported[39].

The percentage of urinary sulphur excreted as inorganic sulphate – in normal subjects about 85% – is lowered in homocystinuria due to cystathionine synthase deficiency as a reflection of the defect in the conversion of methionine sulphur to sulphate[38-41].

### *1.3.4 Tissues*

Only scarce and mostly conflicting data of analyses of tissue amino acids in patients with homocystinuria due to cystathionine synthase deficiency have been reported.
The detection of homocystine in combination with a decreased cystine content was established in liver specimen of three patients[42,43], but not in 5 other studied patients[44,45] whereas in 3 of the latter large concentrations of S-adenosyl-homocysteine were measured[45]. Methionine levels in most of these liver specimen were elevated.
Cystathionine, in normal human brain present at relatively high concentrations[46], was absent or present at very low levels in brains from 3 patients whereas homocystine was not detected and only one specimen showed a moderate elevation of methionine[47,48].
In erythrocytes from one of 4 studied patients homocystine was detected and in 3 of them increased methionine levels were measured[39].
Abnormal high homocystine levels were reported in cultured skin fibroblasts from 3 patients[49], whereas analysis in another patient did not establish either homocystine or elevated methionine concentration[50]. Hair analysis in one patient showed lower cystine concentration than in controls in contrast to normal levels in 3 other patients[4,39,51].

## 1.4 ENZYMATIC CHARACTERISTICS

### *1.4.1 General aspects*

Cystathionine synthase (EC 4.2.1.22), the cytosolic enzyme to start the transsulphuration of homocysteine, is present at relatively high concentrations in human liver[52] and brain[53], and its activity has also been demonstrated in human cultured fibroblasts[54,55], in phytohemagglutinin-stimulated lymphocytes[56], and in long-term cultured lymphocytes[57]. The enzyme purified from human liver is composed of two identical subunits each with a molecular weight of 48,000[52]. Recently however, it has been shown that the primary translational product of the gene for cystathionine synthase in liver tissue as well as in fibroblasts is a subunit of 63,000 molecular weight[58,59]. Posttranslational proteolytic removal of a protein with molecular weight of about 15,000 reduces the size of the original subunits to 48,000 molecular weight. The specific activity of synthase composed of these smaller subunits is much higher than that of the enzyme containing the larger ones, and this is reflected by a 30-fold greater affinity for homocysteine[58]. This proteolytic activation occurs in the liver but not in cultured fibroblasts[59]. Therefore, although the concentrations of cystathionine synthase in liver and cultured fibroblasts are about equal, fibroblast synthase has an about 10-fold lower specific activity than the hepatic enzyme[58].

In a very recent study by Skovby and coworkers[60] it is strongly suggested that the human structural gene for cystathionine synthase is located on chromosome 21. As these authors stated, this would be the first assignment to this chromosome of an enzyme which in homozygous deficient state results in a human disease.

### *1.4.2 Cystathionine synthase deficiency*

After the first report of the detection of the specific defect in cystathionine synthase activity in liver tissue from a homocystinuria patient by Mudd in 1964[6], this finding has been confirmed in other patients. In these liver tissue samples activities of other enzymes involved in sulphur metabolism were in the normal range[6,38,61-69]. Deficiency of cystathionine synthase activity has also been demonstrated in brain tissue of 2 patients[64], in phytohemagglutinin-stimulated lymphocytes in 7 patients[70], and in long-term cultured lymphocytes in one patient[71]. Furthermore from enzyme assays in fibroblast cultures obtained by skin biopsy, deficiency of cystathionine synthase activity has been reported in many more homocystinuria patients[49,54,70-77]. Up to the present more than 600 homocystinuria patients with classical clinical and biochemical features have been diagnosed[7]. In about 200 of them a definitive diagnosis was made by measuring the enzymatic activity in most of them (80%) in cultured fibroblasts of skin biopsies[7]. Residual cystathionine synthase activities in cultured fibroblasts from homozygous homocystinuria patients may range from undetectable to about 15% of mean activity in control subjects[27,49,55,76,78,79].

### *1.4.3 Genetic heterogeneity of mutant cystathionine synthase*

The existence of genetic heterogeneity of the enzyme cystathionine synthase was suspected since it was shown that about 50% of the reported homocystinuria patients did and 50% did not respond to pharmacological doses of pyridoxine, the precursor of cofactor pyridoxal phosphate upon which cystathionine synthase depends for its activity[7,33]. In responsive patients high-dose pyridoxine treatment induces a complete or partial disappearance of their homocystinemia, hypermethioninemia, and homocystinuria, as Barber and Spaeth firstly demonstrated in 1967[80] and as has been confirmed since then by many other authors[27,30,38,39,65-67,81,82]. However, it has to be mentioned here, and it will be discussed further in Chapter 3, that pyridoxine therapy leaves the capacity to handle a major methionine load impaired even in responsive patients[38,67] and that a complete normalization of the pathological amino acid levels is observed only in the fasting state[27]. Pyridoxine responsiveness or non-responsiveness is within sibships constant[7,83], which suggests that there are at least two different genetic mutations in cystathionine synthase.

Using sensitive enzyme assays in liver tissues or cultured skin fibroblasts it has been shown that pyridoxine responsive patients generally had at least traces of residual cystathionine synthase activities albeit as low as 1% to 2% of normal, and that non-responsive patients did not[30,38,55,64,66,68,73,74,78,84]. Nevertheless, immunochemical studies on cultured fibroblasts by Skovby and coworkers[85,86], revealed the presence of immunoreactive cystathionine synthase antigen not only in patients with detectable activity but also in all but one of 6 studied patients with undetectable enzyme activity. Furthermore, they did not find a statistically significant correlation between the percentage residual activity and the percentage immunochemically detected enzyme antigen, indicating that other mechanisms apart from complete or partial gene deletion may account for deficient cystathionine synthase activity.

The correlation between in vivo responsiveness to pyridoxine and the in vitro presence of residual cystathionine synthase activity is not absolute as 9 pyridoxine responsive homocystinuria patients have been reported up to now who did not have detectable residual activities[27,55,76,78,87], and, on the other hand, 6 non-responsive patients who did have such activity in their cultured fibroblasts[55,76,78]. The detection of no residual enzyme activity in the cultured fibroblasts of the former patients could have been the result of insufficiently sensitive enzyme assays and a residual activity may yet be measured in the liver of these patients because of the higher specific activity of the liver enzyme. Nevertheless, the finding of detectable enzyme activities in pyridoxine non-responders indicates that the mere presence of residual cystathionine synthase activity is not the only prerequisite for clinical pyridoxine responsiveness.

Moreover, it has been shown that enhancement of residual activity of cystathionine synthase in cultured fibroblasts by addition of 1mM pyridoxal phosphate to the cell extracts may occur to a similar or even much higher extent (up to 1000%) as in control cell lines (30% to 50%)[55,76]. Neither the basal enzyme activity nor its increase in vitro by added pyridoxal phosphate was correlated with the in vivo responsiveness upon pyridoxine treatment. In fact, the higher responses in vitro even were observed in some of the exceptional clinically non-responsive patients with detected residual activity[55,76].

It has been suggested that some mutant cystathionine synthases have a reduced affinity for pyridoxal phosphate and thereby less thermostability than control synthase[30,76,84,88], which condition may be overcome by pyridoxal phosphate in excess[69,76,84]. However, this is not in line with the above mentioned observed pattern of stimulation of residual enzyme activity in vitro by added pyridoxal phosphate. In fact, the more marked stimulation was found most often in clinically non-responders and the enhancement in most pyridoxine responsive patients did not exceed that in controls. Furthermore, studies on thermostability of mutant cystathionine synthases extracted from fibroblasts of pyridoxine responsive patients did not reveal any consistent pattern because the enzyme of some patients showed less[75,76], and some equal[38] or even greater stability[75] than the enzyme of control subjects.

These findings together lead to the conclusion that in homocystinuria due to cystathionine synthase deficiency heterogeneity abounds. It is not unlikely that in the future after extended studies it will be demonstrated that as much different genetic determinants affecting cystathionine synthase exist as there are different sibships with cystathionine synthase deficiency[83].

## 1.5 DIAGNOSIS

The finding of undetectable activity of cystathionine synthase or lowered activity up to about 15% of the mean control level, measured in cultured fibroblasts from a skin biopsy[27,49,55,76,78,79,87], provides the final diagnosis of homocystinuria due to ***homozygous*** cystathionine synthase deficiency. The rather sophisticated procedure of cell culturing and enzyme assay, which techniques have been reported elsewhere[76,87], is reserved for patients with one or more classical clinical features of the syndrome in whom quantitative amino acid analysis in serum or plasma by column chromatography established the biochemical characteristics of the homozygous state of cystathionine synthase deficiency. Details of the amino acid analysis by column chromatography have been described in the "Methods" section of Chapter 6.

The presence of homocystine even in fasting serum or plasma in association with elevated levels of homocysteine-cysteine mixed disulphide and in combination, almost consistently, with hypermethioninemia and, usually, with lowered cystine concentrations constitutes the biochemical diagnosis of cystathionine synthase deficiency in the homozygous condition. In the patients showing homocystinemia *without* hypermethioninemia the results from the enzyme assays allow differentiating between cystathionine synthase deficiency – 6% of these patients are not hypermethioninemic[7] – and deficiencies of the re-methylation of homocysteine to methionine due to the rare genetic conditions of 5,10-methylenetetrahydrofolate reductase defect or deficient formation of cobalamin cofactors, as discussed in Chapter 1.3.2. Such a re-methylation deficiency can also be caused by the familially occurring defective absorption of vitamin $B_{12}$ in the presence of intrinsic factor (Imerslund's syndrome)[89,90], or by extreme nutritional vitamin $B_{12}$ deficiency in children[91].

The result of measurement of amino acid concentrations by column chromatography in urine is less indicative than in blood. Admittedly, homocystinuria is a consistent finding in cystathionine synthase deficiency as stated in Chapter 1.3.3, but only in the case of concomitant elevated urinary methionine levels it is conclusive of cystathionine synthase deficiency. However, hypermethioninuria is not altogether obligatory in cystathionine synthase deficiency and it is lacking in cases with re-methylation defects. Moreover, pseudohomocystinuria without hypermethioninemia can occur when in case of excessive excretion of cystathionine bacterial conversion in the contaminated urine produces homocystine from cystathionine[92,93]. Suchlike cystathioninuria is present not only in γ-cystathionase deficiency, but has also been reported as a transient feature in normal infants especially newborns[94-97], in vitamin $B_6$ deficiency[98-101], in severe liver damage[102-105], in thyrotoxicosis[106], and in some patients with neuro- or ganglioblastoma[107,108] and hepatoblastoma[103,109-111].

Semiquantitative methods to screen for the presence of homocystine and elevated levels of methionine in blood and urine are too insensitive to make the biochemical diagnosis in all cases of homocystinuria. Paper or thin-layer chromatography, high-voltage paper electrophoresis, and two-way sequential paper chromatography[34] therefore may present false negative results. The Guthrie bacterial assay for the detection of hypermethioninemia[112], regularly used in newborn screening programs, is an inadequate method both due to its low sensitivity (positive test only at levels of methionine exceeding 100 μmol/l) and its failure to detect those homocystinuria patients without hypermethioninemia. Furthermore, especially in newborns the use of this assay is misleading because methionine in the blood of cystathionine synthase deficient newborns is rising to elevated levels during the first weeks of life and cases might be missed soon after birth when routine screening is usually performed[113,115].

In many earlier studies the qualitative cyanide-nitroprusside reaction for the detection of homocystine in urine, the so-called Brand test[116], was used. Because of its low sensitivity this classical test gives false negative results in up to 30% of homocystinuria patients[77]. Due to a positive reaction in acetonuria and even in case of high urinary creatinine concentration[117], false positive results may be found. Moreover, a positive test can also be established in cystinuria.

Prenatal diagnosis of homocystinuria in fetuses at risk has been performed since it appeared possible to measure cystathionine synthase activity also in cultured amnion fluid cells[44,54,72,118,119]. The obligate heterozygous state of fetuses carried by homozygous mothers has been reliably predicted in several cases[44,72,120]. In 5 fetuses at risk because of affected older sibs in their families, the prenatal determination of enzyme activity correctly diagnosed heterozygosity in 2[54,118], the normal condition in 2[119,121], and the homozygous state in one of them[121]. Therefore, homocystinuria due to cystathionine synthase deficiency is one of the inherited metabolic diseases which can be reliably diagnosed in utero.

## 1.6 CLINICAL FEATURES

In homozygous homocystinuria four major organ systems predominantly are involved: the eye, the skeleton, the vascular and the central nervous system. Clinical abnormalities are given in Table 1 and will be discussed in some detail below.

It is of note that in homocystinuria patients clinical abnormalities generally are absent at birth and progressive features develop in childhood and adolescence[7,122]. The clinical picture is extremely heterogeneous ranging from patients presenting all of the complications to individuals with no overt clinical involvement[36,77,123-125]. The clinical expression among affected sibs within one family has been reported to be more similar[122,126,127], except in one study[77]. The wide variation in clinical manifestations might be the reflection of the abound heterogeneity of mutant cystathionine synthase as discussed in Chapter 1.4.3.

### *1.6.1. Ophthalmological manifestations*

The most consistent clinical finding in homocystinuria is ectopia lentis (Chapter 4, Figure 3). In a recent survey of about 600 patients, Mudd et al.[7] calculated that this complication, occurring not before the age of 2 years, is present in 50% of untreated pyridoxine responsive patients at age of 10 and in 50% of untreated non-responsive patients at age of 6 years. The probability in an untreated patient to have no ectopia lentis by age of about 40 years is as low as 3%. From the patients diagnosed up to now 82% presented this complication at some age[7].

Ectopia lentis in homocystinuria, bilaterally and usually downwards, is caused by disruption of thickened zonulae fibres[128,129]. It is of interest that in Marfan's syndrome regularly the lenses are dislocated upwards and in this disorder the zonulae fibres are thin, elongated and not disrupted[125,130-132]. The disruption of the zonulae fibres in homocystinuria leads to increased curvature of the lens and thereby to lenticular myopia and astigmatism, resulting in reduction of visual acuity. Iridodonesis, the quivering of the iris by moving the eye ball, is a sign of the loosening of the lens. Dislocation of the luxated lens into the anterior chamber causing acute pupillary block glaucoma accompanied by severe eye pain (Chapter 4, Figure 2), may be the presenting symptom in homocystinuria patients[125,133-136].
Retinal detachment may occur spontaneously or as a complication of lens surgery[131]. Occlusion of the central retinal artery either primary[137] or secondary to acute glaucoma[138] has been reported. Optic atrophy has been observed, usually on the basis of glaucoma[139,140], or due to occlusion of the retinal artery[138].

Ectopia lentis is a characteristic finding not only in homocystinuria and

*Table 1.* Clinical abnormalities*

| | | |
|---|---|---|
| I | EYE | ECTOPIA LENTIS |
| | | MYOPIA AND ASTIGMATISM |
| | | IRIDODONESIS |
| | | ACUTE PUPILLARY BLOCK GLAUCOMA |
| | | retinal detachment |
| | | retinal artery occlusion |
| | | optic atrophy |
| II | SKELETON | OSTEOPOROSIS |
| | | SCOLIOSIS |
| | | "FISH-VERTEBRAE" |
| | | MARFANOID FEATURES: |
| | | DOLICHOSTENOMELIA |
| | | DECREASED UPPER/LOWER SEGMENT |
| | | ARACHNODACTYLY |
| | | EXTREME HEIGHT |
| | | genu valgum |
| | | pectus carinatum or excavatum |
| | | high-arched palate |
| | | protrusion of upper teeth |
| | | pes cavus or planus |
| | | abnormal bone age |
| | | "Charlie Chaplin"-gait |
| III | VASCULAR SYSTEM | PREMATURE ARTERIOSCLEROSIS |
| | | ARTERIAL AND VENOUS THROMBOEMBOLISM |
| | | malar flush and livedo reticularis |
| IV | CENTRAL NERVOUS SYSTEM | MENTAL RETARDATION |
| | | CONVULSIONS |
| | | ABNORMAL ELECTROENCEPHALOGRAM |
| | | psychiatric disorders |
| V | MISCELLANEOUS features | fair and thin skin |
| | | fair and brittle hair |
| | | light and blue iris |
| | | inguinal hernia |
| | | fatty liver and hepatomegaly |

*frequent abnormalities in capitals

Marfan's syndrome, but it has also been reported in inherited metabolic conditions as hyperlysinemia[141] and sulphite oxidase deficiency (Chapter 1.2, Figure 1)[142-144], in Weill-Marchesani syndrome and in a number of ocular disorders[145].

### *1.6.2 Skeletal abnormalities*

Osteoporosis, especially of the spine, is a very conspicuous roentgenographic finding in homocystinuria. Although not present at birth[146], it has been observed as early as at age of one year[122,147]. In several series the frequency has been reported as high as 90% to 100% of studied patients[27,147-150], with one exception in which study the frequency was only 35%[151]. Mudd, surveying about 600 patients diagnosed in various countries[7], concluded that 50% of the untreated patients showed radiographic spinal osteoporosis at age of 17 years (pyridoxine non-responsive patients at age of 12 and responsive patients at age of 20 years). The probability in a patient at age of 30 years not to have osteoporosis could be calculated as low as 20%[7]. Osteoporosis may cause vertebral collapse, pathological fractures, and spondylolisthesis[147,152-154]. Scoliosis or kyphoscoliosis is reported in up to 50% of cases[27,125,147,148,150,151,155,156], and is generally held to be caused by osteoporosis. It can be progressive and may lead to pulmonary complications (Figure 1; Chapter 4, Figure 1)[125]. The vertebrae often are biconcavely flattened presenting the roentgenographic sign of "fish-vertebrae" (Figure 2).

Marfanoid features have been reported in about 30% of the patients[27,125,146-148,150,151,156] such as thin and excessively lengthened long bones – a condition known as dolichostenomelia (Chapter 4, Figure 4) –, increased arm span, decreased ratio upper (vertex to pubis) segment to lower (pubis to heel) segment, arachnodactyly with abnormal metacarpal index (Chapter 2.6.2, Figure 1) and extreme height. The latter two characteristics have been observed each in about 50% of the patients[27,124,125,150,155].

Several minor symptoms have been regularly established as there are genu valgum, pectus carinatum more than excavatum, high-arched palatum, protrusion of the upper teeth, pes cavus and exceptionally pes planus[150]. Bone age is usually normal, but has occasionally reported to be retarded or accelerated[147,148,153]. In many patients a typical "Charlie Chaplin"-gait has been described[152,157].

### *1.6.3 Vascular pathology*

Premature arteriosclerosis and arterial and venous thromboembolism are life-threatening complications in homocystinuria. Mudd et al. in 1985 calculated from data of 629 homocystinuria patients a probability of 30% in untreated patients to suffer from thromboembolism before the age of 20 and 50% before the age of 30 years, with only minor differences in these probabilities between pyridoxine responsive and non-responsive patients[7]. In 158 of these 629 patients a total of 253 thromboembolic events occurred, 51% in peripheral veins, 32% were cerebrovascular accidents, 11% affected peripheral arteries and only 4% led to myocardial infarctions.

Vascular occlusions in homocystinuria patients may occur in venous and arterial vessels at any age. Thrombophlebitis, even in the first decade of life[117,158], is the most frequent complication particularly in the legs and then often leading to fatal pulmonary artery emboli[5,27,122,149,152,159] or chronic corpulmonale[160,161]. Thrombosis of the femoral veins[162,163], the vena cava inferior and iliac veins[117,152,164,165], the renal veins[122] and vena portae[152] has been reported.

Cerebral thrombosis occurred in some patients already before the age of 2

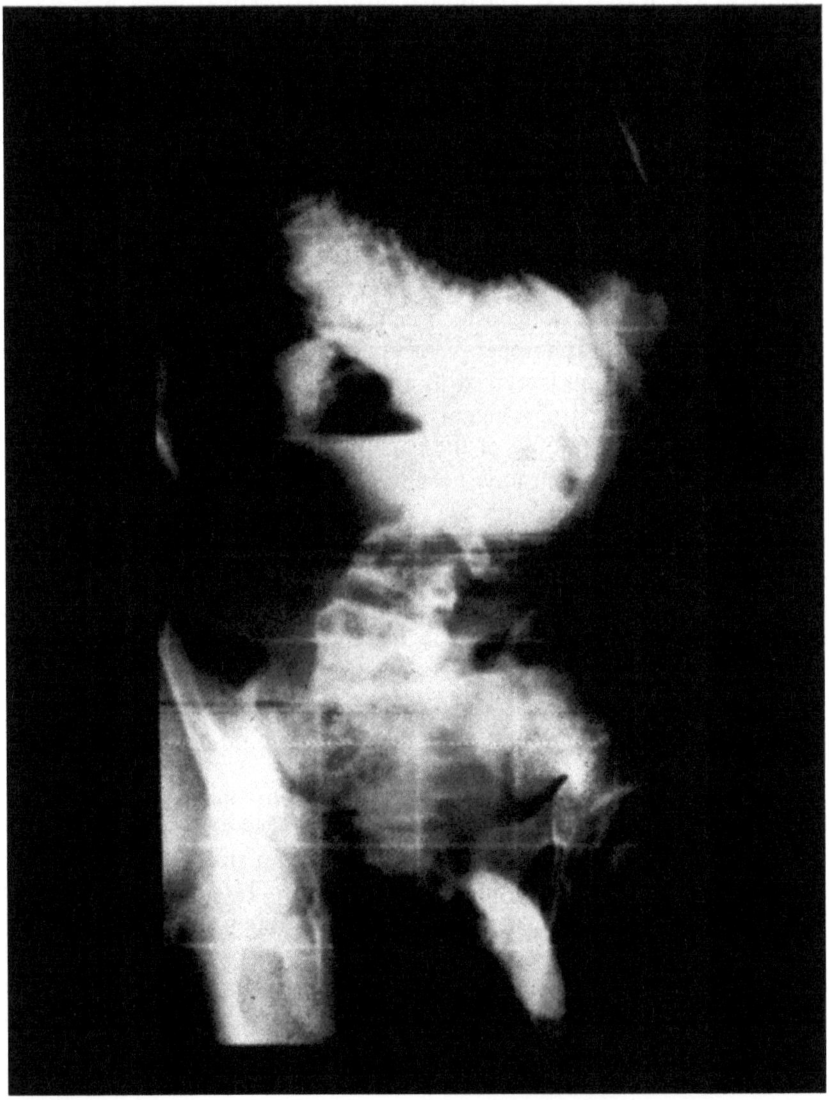

*Fig. 1* Severe scoliosis in a homozygous homocystinuria patient (patient no 16, Chapter 2), leading to pulmonary complications.

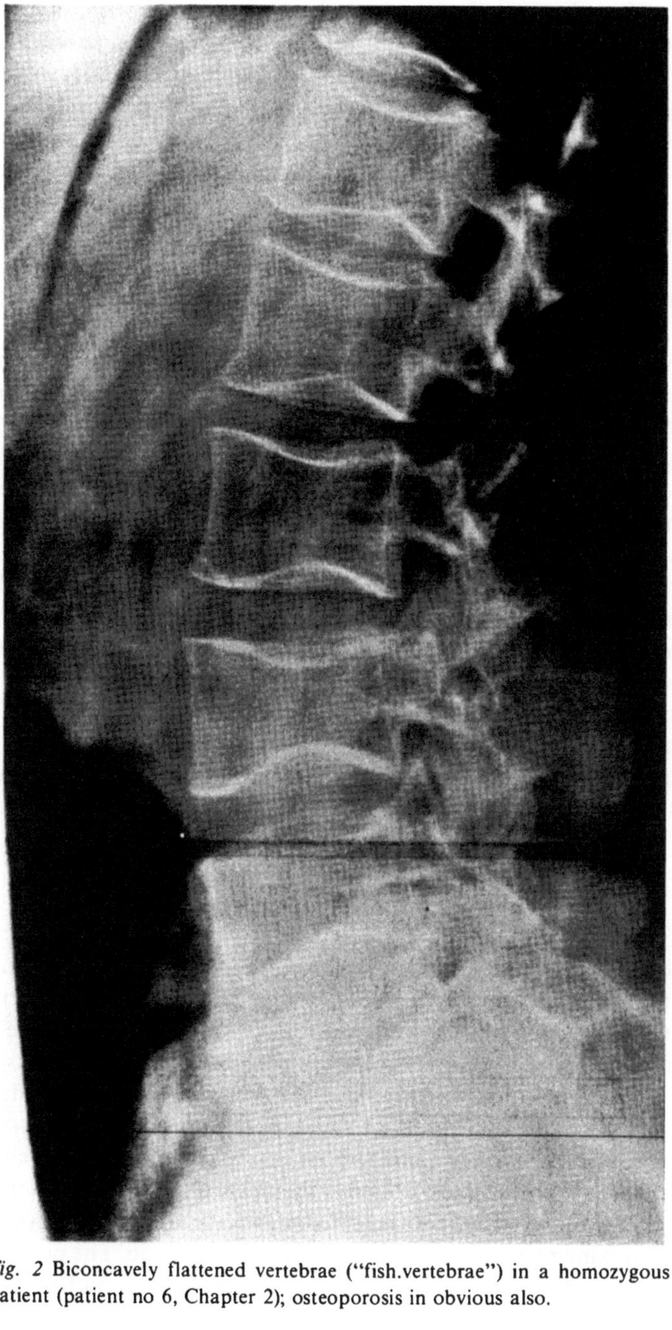

*Fig. 2* Biconcavely flattened vertebrae ("fish.vertebrae") in a homozygous homocystinuria patient (patient no 6, Chapter 2); osteoporosis in obvious also.

years[122,123,159] and cerebrovascular accidents due to occlusions of the carotid arteries or intracerebral arteries are frequently observed, even in adolescence[27,117,125,127,136,166]. Peripheral arteries may be prematurely affected by arteriosclerotic lesions and thrombotic occlusions, as has been observed in the aorta, the femoral and the iliac arteries[117,148,160,167,168], and the mesenteric arteries[117,127,160,162,164]. Occlusion of the renal arteries often results in hypertension[125,127,136,160,162]. Coronary artery occlusion leading to myocardial infarction has been observed in an 11 year old patient[160] and in several patients before the age of 20[117,152]. However, of all thromboembolic events reported in the group of 629 patients by Mudd[7] only 4% produced myocardial infarctions.

In earlier studies it was stated that in homocystinuria patients anaesthesia, surgery and long-standing immobilization should be withheld in view of their extremely high risk of thromboembolic complications[157,169,170]. If absolutely indicated, surgery should be performed under special precautions to prevent hypotension, dehydration and stasis of the blood, and during strict anticoagulant treatment[125,170,173]. Mudd however, recently surveying 586 major surgeries and eye operations in homocystinurics, established not more than 25 postoperative thromboembolic complications, fatal in only 6 patients[7]. It seems therefore that the risk to develop thromboembolic events after surgery in homocystinurics, although present, has been overestimated in the past due to dramatic reports on only few patients.

Dilatation or coarctation of the arteriosclerotic aorta[117,122,124,152,160-162,167-169], bundle branch block[174], and aortic valve incompetence[125,152] have been reported as vascular complications in homocystinuria. Similarly, vascular pathology has been accounted for the malar flush and livedo reticularis which are strikingly present in a number of patients[117,152].

### *1.6.4 Central nervous system involvement*

Mental retardation is the most frequent sign of involvement of the central nervous system in homocystinuria. Under the age of 2 years it may present itself as delayed motor and speech development[5,126,151] and at ages of 4 to 6 years the impairment of mental capacity becomes overt. In most of the cases it is progressive[157] and even an already obtained level of development may deteriorate[151]. In the survey by Mudd[7] it is shown that 61% of the patients diagnosed up to now were severely or mildly retarded, 38% of the pyridoxine responsive and 85% of the non-responsive patients. The IQ's for the patients in whom these data were available, ranged from 10 to 138 with a median of about 64, in pyridoxine responders 78 and in non-responders 56. The percentages of patients with IQ's of 90 or more were significantly different between the responders and non-responders, 22% versus 4%, indicating that pyridoxine responsive homocystinuria patients are less susceptible to mental capacity impairment than non-responding patients.

Convulsions are reported to occur in 21% of all patients diagnosed up to now[7]. The incidence in pyridoxine responsive patients was 17%, in non-responsive ones 23%. About 70% of the observed convulsions were grand mal seizures.

Electroencephalographic abnormalities have been established in many patients, also in those without a history of convulsions[27,68,122,124,126,160,175]. Excessive slow wave activity is the most frequent observation[122,124,126,175], but also discharges as spikes or sharp waves have been found[68,122]. On the other hand, normal electroencephalograms have been established in several patients with histories of convulsions[4,163,176].

Other neurologic abnormalities incidentally described in homocystinuria patients such as extrapyramidal rigidity[122,157,163,177-179], muscular hypotonia[122], hyperactive deep-tendon reflexes and a postive Babinski sign[4,176], have been most probably associated with minute cerebrovascular infarctions in specific brain structures[83,122,162,176]. Therefore, they are not to be classified as typical signs of homocystinuria.

Psychiatric disorders have been reported in a number of homocystinuria patients. Schizophrenia is observed[135,151,152,160], as well as psychotic depression[27], hyperkinesis with manic episodes[160,180], hysteria[181], and "nervous breakdown"[133]. However, because of the high frequency of mental disturbances noted also in non-homocystinuric family members of several of these homocystinuria patients[4,122,152], it is not altogether excluded that the reported psychiatric illnesses among these homocystinurics reflect only chance associations[83].

### *1.6.5 Miscellaneous features*

A fair, thin skin with increased fragility and fair, brittle hair were striking features in some reported patients[26,122,152,157,160,180]. Furthermore, it is stated that in 75% of homocystinuria patients irides are light or blue[130]. The incidence of inguinal hernia is increased among homocystinurics[5,124,161]. In a number of patients hepatomegaly due to fatty infiltration was established[160,162,164,182].

## 1.7 PATHOPHYSIOLOGY

The mechanisms by which the basic defect of cystathionine synthase comes to its specific clinical expression in homocystinuria, still are largely obscure. Toxicity of accumulated homocysteine is generally held to be the key factor relevant to the pathophysiology[33]. In addition, several of the clinical abnormalities may originate from deficiences of substances distal to the metabolic block such as cystathionine, cysteine and sulphate (Chapter 1.2, Figure 1). The pathophysiological concepts for which sufficient experimental support is available, are discussed below.

### *1.7.1 Structural collagen abnormality*

There is ample evidence that homocysteine in excess inhibits the normal cross-linkage of collagen due to binding of its sulphhydryl group with the aldehydic group in collagen monomers, thereby interfering with the formation of intermolecular cross-links in collagen[117,183-185]. In collagen from skin specimen of some homocystinuria patients a higher percentage of acidic-soluble collagen was established than in control collagen[186,187]. Moreover, the ratio of single to double chain components was found to be abnormally high in collagen from homocystinurics[186]. This structural collagen abnormality in homocystinuria might account for the origin of some of the clinical abnormalities typical for the syndrome such as osteoporosis, deformities of the vertebrae, the spine and the long bones, the fragility of the thin skin, and the increased incidence of inguinal hernias.

The ocular zonulae fibres, composed of non-collagenous glycoproteins with high content of cysteine[188] are postulated to disrupt, with ectopia lentis as a result, due to lack of cystine disulphide bonds between the glycoproteins[139,189] as well as because of disjunction of the cystine disulphide bonds after disulphide exchange with homocysteine[142].

### *1.7.2 Homocysteine-induced endothelium injury*

The finding of such high incidence of thromboembolic events affecting homocystinuria patients both in veins and arteries, has attracted much attention to the possibility of abnormalities in platelet functions or blood clotting in patients as a cause of the thrombotic tendency. However, up to the present only contradictory data on such abnormalities have been reported in homocystinuria patients. Abnormal platelet adhesiveness was established in homocystinuria patients by some investigators[39,82,161,164,190], but not by others[152,191-194]. Platelet aggregation has been found to be abnormal in some patients[195,196], not in many others[33,197]. Data on morphological abnormalities of platelets from homocystinuria patients, detected by electron microscopy[149], have not been confirmed in an other report[198]. Incidental findings of abnormal activation of the Hageman factor by homocystine[199], antithrombin-III deficiency in 2 patients[200], and proconvertin deficiency in 9 homocystinurics[201-202], have not followed up by similar findings in further studies.

Pathological examination of venous walls from homocystinuria patients never did have shown any abnormality[203], except for fibrous endophlebitis in a renal vein from a single patient[162]. On the contrary, striking arteriosclerotic lesions have been observed at autopsy of many patients in their large, medium, and small arteries. These lesions, multiple, patchy or arounding and narrowing the vascular wall, were established as intimal fibrosis, a thickened lamina elastica interna with cracks due to frayed and discontinuous elastic

fibres, and a thickened media with proliferation of split muscle fibres and interstitial collagen[117,136,152,160,162,166,167,176,203]. Suchlike lesions had also been found in patients with defects in the re-methylation of homocysteine to methionine showing homocystinemia without hypermethioninemia and lowered cystine levels (Chapter 1.3.2), and, therefore, the attention was focused at homocysteine as a vasotoxic agent[83].
Animal experiments and in vitro studies on cultured human endothelial cells, indeed, have given support to the hypothesis that homocysteine in excess induces endothelial cell injury which initiates this process of premature arteriosclerosis. Parenteral administration of homocysteine in rabbits produced focal arteriosclerotic plaques similar to those in homocystinuria patients and in early human arteriosclerosis[204,205]. After continuous homocysteine infusions during 3 months in baboons, Harker and coworkers established arteriosclerotic and preatherosclerotic intimal lesions in all of these animals[194,206-208]. Areas of patchy endothelial cell loss at which platelets were found to adhese, were observed already after one week of homocysteine infusion[194]. A linear relationship was calculated between the induced homocystine concentrations in plasma and the observed endothelial cell loss, the platelet turnover and the score of the intimal arteriosclerotic lesions in these animals. Treatment with dipyridamole, a platelet aggregation inhibiting agent, prevented most of the platelet utilization and arteriosclerotic lesions without altering the amount of desquamated endothelium[194,206,207]. Harker established also a 50% shortening of platelet survival time in homocystinuria patients, and treatment with dipyridamole, or pyridoxine in the case of pyridoxine responsive homocystinurics, normalized the platelet survival[194]. Admittedly, this enhanced platelet consumption in homocystinuria patients was not confirmed by other authors[198,209], but methodologic differences between the studies have been accounted for this discrepancy[210,211]. Homocysteine at concentrations comparable with those in plasma of untreated homocystinurics, resulted in an in vitro study in injury of cultured endothelial cells in terms of detachment of cells and abnormally high $^{51}Cr$ release from labelled cell monolayers, proportional to the homocysteine concentration in the medium[212]. In a recent study on cultured endothelial cells from an obligate heterozygote for homocystinuria, it was shown that cystathionine synthase was present in endothelium too[213]. It seems therefore likely that endothelial cells, if deficient in this enzyme activity as in homocystinuria, contribute to their own dysfunction. Indeed, the cultured endothelial cells even in this heterozygote proved to be more susceptible to homocysteine-mediated injury than control cells.
These findings together support the response-to-injury hypothesis in homocystinuria, proposing that homocysteine-induced endothelial cell injury leads to platelet-mediated migration and proliferation of smooth muscle cells, elaboration of extracellular connective tissue matrix, and finally the development of typical arteriosclerotic lesions[214].

### *1.7.3 Central nervous system dysfunction and homocystinuria*

Homocysteine has been shown to be a potent excitant of cortical neurons[215,216] and convulsive effects of intraperitoneal injections of homocysteine in animal experiments have been described in several reports[217-221]. Cerebral metabolic changes accompanying homocysteine-induced seizures have been studied extensively[219,220-224]. Recent results from animal experiments show that inhibition of the postsynaptic γ-amino-butyric-acid (GABA) receptor by homocysteine is a possible explanation for its seizure activity[224]. However, the actual mechanisms of seizure induction in both humans and experimental animals remain largely obscure[224].

At autopsy brains of homocystinuria patients proved to be extremely cystathionine-deficient, whereas this amino acid is present in normal brain tissue in high concentrations (Chapter 1.3.4). Therefrom it has been reported that mental deficiency and neurologic dysfunction in homocystinuric patients might be due to absence of cystathionine in their brains[47,48]. However, indications that normal brain function is not in need of cystathionine, contradict this hypothesis[83].

## 1.8 MODE OF INHERITANCE

Pedigree studies pointed out an autosomal recessive inheritance in homocystinuria[117,124]. In concordance with this trait is the observation of intermediate activities of cystathionine synthase in cultured fibroblasts in both parents of homocystinuria patients (Chapter 7).

Mutant cystathionine synthase is held originating most frequently from a single base change in the structural gene leading to substitution of a single amino acid in the enzyme[86]. Due to the composition of the enzyme cystathionine synthase of two subunits (Chapter 1.4.1), a homozygote for homocystinuria in fact might be a compound heterozygote, i.e., in each of the subunits a different mutation is expressed. Recently, suchlike compound heterozygosity was firstly documented in a homocystinuria patient[86]. In view of the supposed large heterogeneity in cystathionine synthase mutants (Chapter 1.4.3), it is intelligible to assume that the majority of homozygous patients indeed are compound heterozygotes, as is likely to be confirmed by future studies at the level of the subunits of enzyme mutants and at the gene level using recombinant DNA techniques[60,86].

## 1.9 INCIDENCE

Newborn screening programs for the detection of various inborn errors of metabolism have been conducted for the last 10 years or longer in the United States, Australia, New Zealand, and in many European countries. In the majority of these programs screening for homocystinuria has been included. Among a total of more than 19 million screened newborns, the incidence for homozygous homocystinuria appeared to be 1 in 200,000[83]. In several areas the incidence was found to be much higher: 1 in 57,000 in Ireland[225], 1 in 73,000 in Northern Ireland[225], 1 in 85,000 in the region of Manchester[226].

In most of the screening programs the Guthrie assay for hypermethioninemia has been used, generally in the first 2 weeks of life, in some paper chromatography of blood[226] or urine[123]. The possibility to miss cases of homocystinuria using such semiquantitative methods, especially if performed in the first weeks after birth, has been pointed out in Chapter 1.5. Indeed, several homocystinuria patients have been reportedly missed in current newborn screening[113,115,124]. Particularly in newborn pyridoxine responsive homocystinuria patients this risk is present because these patients are more regularly not hypermethioninemic and less homocystinuric during their first weeks of life than non-responsive patients[7,83]. Therefore, the true incidence of homocystinuria is likely higher than 1 in 200,000 as established by newborn screening. Based upon the mutation rate formula, assuming a reproductive fittness of 0.1 compared to 1.0 in the average population, McKusick calculated that the frequency of homozygous homocystinuria should be as high as 1 in 45,000[117]. This calculation seems to be in agreement with the estimated incidence of 1 in 24,000 for Northern Ireland[227] and 1 in 58,000 for New South Wales, Australia[123], both based upon the number of diagnosed homocystinuria patients in the respective populations.

The frequency of homozygous homocystinuria established among patients in institutions for the mentally retarded, is reportedly as high as 1 in 300 to 4,500[157,192,228]. Among individuals with non-traumatic ectopia lentis a frequency of 1 in 20 has been claimed[152].

## 1.10 TREATMENT

### *1.10.1 Pyridoxine*

About 50% of the reported patients with homocystinuria due to deficiency of cystathionine synthase, a pyridoxal phosphate dependent enzyme, respond to treatment with pyridoxine in pharmacological dose[7]. Such response is not due to overcoming of a depletion or defective metabolism of vitamin $B_6$ in these patients[83]. The postulated mechanisms of the pyridoxine effect at the enzyme level have been discussed in Chapter 1.4.3. It has also been pointed out in that section that responsiveness is correlated – albeit not absolutely – with the presence of detectable residual cystathionine synthase activity in cultured fibroblasts: pyridoxine responders show at least traces of enzyme activity, non-responders do not. The presence of some residual activity in responsive patients still enables them to convert about 1 gram methionine into inorganic sulphate per day, in contrast to non-responders who lack this capacity[38]. Pyridoxine treatment enhances about twofold this capacity in responsive patients. However, normal humans are able to metabolize at least 12 grams of methionine a day. Therefore, complete restoration of the abnormal amino acid levels, as has been reported in many responsive patients while on treatment with pyridoxine[30,38,39,65-67,80-82,154], may be observed only in the fasting state or in case of moderate protein intake. Profiles of relevant amino acid levels in the fasting state and after a standardized methionine load off and on pyridoxine treatment are given and discussed extensively in Chapter 3. From these data it is obvious that pyridoxine therapy leaves the capacity of pyridoxine responsive patients to handle a major methionine intake essentially unchanged.

A homocystinuria patient should not be considered pyridoxine non-responsive unless daily doses of pyridoxine up to 1000 mg have been administered for at least 6 weeks[229]. The most regular prescribed dose is 750 mg daily in adults and at least 150 mg daily in children. In exceptional patients a maximal biochemical response is already achieved at lower doses[65,83].

Enhanced re-methylation of homocysteine to methionine (Chapter 1.2, Figure 1) may lead to folate depletion in homocystinuria patients[229,230]. This can be the cause of spurious non-responsiveness in a patient and, therefore, the responsiveness to pyridoxine should be interpretated only in the presence of normal folate levels. While on treatment with pyridoxine, folate deficiency for unknown reason can be exacerbated leading to a decrease of the primary pyridoxine effect[82,229]. Thus, supplemental folic acid therapy is often required in homocystinurics.

No side effects of high-dose pyridoxine administration have been reported in homocystinuria patients up to the present. Recently, ataxia and sensory neuropathy have been observed in 7 non-homocystinuric adults on the misuse of megadoses as high as 2 to 6 grams of pyridoxine daily for 2 to 40 months[231]. In one case such side effects were established while taking only 500 mg daily[232]. All persons improved after withdrawal from pyridoxine. Among about 300 responsive homocystinuria patients these adverse effects of pyridoxine therapy never have been reported since the first treatment was started in 1967.

### *1.10.2 Diet*

A methionine-restricted diet to correct the biochemical abnormalities is used in the treatment of pyridoxine non-responsive homocystinuria patients. In 1966 firstly Komrower reported the outcome of a homocystinuria patient on such diet started in the second week of life and used throughout infancy[181,229]. This diet was gelatin-based with supplementation of essential amino acids except for methionine and with addition of cystine. To maintain normal

growth a methionine intake was required of about 30 mg/kg body weight-/day in the first year of life, 20 mg in the second and from the third 10 to 15 mg. Under this regimen methionine in the blood was near normal and homocystine mostly not detectable, but cystine levels remained low[181].
The gelatin- or, alternatively, soybean[230]-based diets largely have been superseded by diets based on methionine-free L-amino acid mixtures, containing also minerals and other nutrients. Carbohydrates, fat, and vitamins have to be added to these commercially available formulas. The ingestion of small, controlled quantities of milk or low-protein food in addition is permitted under the proviso of maintaining methionine levels near the normal range and the presence of only minimal homocystinemia. A cystine supplementation of 150 to 200 mg/kg body weight is daily required in homocystinurics who lack any residual cystathionine synthase activity, to achieve at least low-normal blood concentrations[22]. Nevertheless, in some cases it appeared to be impossible to raise blood levels of cystine even with high oral supplements[31]. Cysteine administration however, was able to restore the blood level of cystine in a patient to a much higher concentration than cystine ingestion, possible due to its faster rate of intestinal absorption[231].
Unfortunately, despite of the amelioration of the biochemical abnormalities these dietary regimens have regularly to be stopped especially in older homocystinuric infants because of lack of compliance[164,232].

In several pyridoxine responsive homocystinuria patients pyridoxine therapy does not induce a normal or near normal profile of amino acids in the fasting state. In these so-called partially responsive[154] patients pyridoxine treatment should be combined with a dietary methionine restriction, albeit not necessarily as strict as in non-responsive homocystinurics. Even in "completely responsive" patients major methionine loads from the diet should be avoided and a more equal distribution of the methionine intake during the day is mandatory, as is discussed in Chapter 3.

### *1.10.3 Additional therapeutic measures*

Enhancement of re-methylation of accumulated homocysteine to methionine, in addition to restricted methionine intake in pyridoxine non- or "partially" responsive homocystinurics, may induce a further decrease of homocysteinemia (Chapter 1.2, Figure 1). High-dose folate administration even in the case of no folate depletion, has been used for this purpose[65,233,234].
Therapy with choline[31] or betaine increases the rate of re-methylation through betaine-homocysteine methyltransferase (Chapter 1.2, Figure 1). The results of betaine supplementation in a dose of 6 to 10 grams daily have been reported in about 17 pyridoxine non-responsive patients[164,181,235-237].
The accumulation of homocysteine substantially decreased whereas cystine levels increased, but also methionine concentrations rose as a result of the

enhanced re-methylation to methionine. Long-term treatment showed a beneficial clinical effect in these patients and no side effects of betaine or toxicity of the more marked hypermethioninemia have been observed.

In 3 pyridoxine non-responsive patients D-penicillamine in a dose ranging from 3.5 to 5 grams daily reduced largely homocysteine accumulation without changing methionine and cystine levels[25]. D-penicillamine, a sulphhydryl containing agent, binds free and protein-bound homocysteine to form homocysteine-D-penicillamine disulphide, which is excreted in the urine. No undesirable effects of this treatment have been observed during 2 months of therapy in these homocystinurics. Nevertheless, the usefulness of this regimen has yet to be established as even moderate doses of D-penicillamine up to 750 mg daily are regularly causing severe side effects in the treatment of various disorders such as reumatoid arthritis, Wilson's disease, and cystinuria.

In addition to measures to reverse the biochemical abnormalities in homocystinuria, several regimens to prevent thromboembolic complications have been recommended. In one study every patient, although "completely" pyridoxine responsive, was treated with a combination of pyridoxine and a coumarin derivative[77] (Chapter 2.5). Harker observed a normalization of decreased platelet survival and a marked reduction of intimal lesions by pyridoxine therapy in pyridoxine responsive patients and by dipyridamole, a platelet aggregation inhibiting agent, in non-responsive ones[194,206,207,210,214]. From these data he recommended treatment with dipyridamole, 400 mg daily, or with the combination of dipyridamole, 100 mg daily, and acetylsalicylic acid, 1 gram daily, in pyridoxine non-responsive patients. In 2 such patients no thromboembolic events occurred during 3 years under this regimen whereas in the year before the start of this medication at least 5 such complications had been established in them. In contrast, in 2 further non-responsive patients this therapy did not prevent thromboembolic episodes[238,239].

Pregnancy as an additional risk of thromboembolism has been discouraged in homocystinuria females[240]. However, in several pyridoxine responsive females no complications in pregnancy and a normal outcome have been reported after control of the biochemical abnormalities during pregnancy by pyridoxine and additional folate treatment[7,213,241]. Nevertheless, the results might be different in non-responsive females with high homocysteinemia during pregnancy[7,241].

Drugs with an increased risk of thromboembolism, such as oral contraceptives, should be avoided in homocystinuria females[117,125,136,240,242]. Apart from their additional thrombotic risk, estrogenic steroids may induce pyridoxine and folate depletion[243], resulting in further derangement of methionine metabolism. In homocystinuric females the occurrence of major vascular and even fatal complications have been reported shortly after the start with oral contraceptives[125,136,242].

1.11 PROGNOSIS AND THE CLINICAL EFFECT OF TREATMENT

In 1972 McKusick concluded, based upon survival data in 83 patients with homocystinuria, that approximately 50% of them had died before the age of 20 years, and 75% by age of 30[117]. Mudd in 1985 surveyed the survival data in more than 600 patients and established a far better prognosis quoad vitam: 11% of the patients had died before the age of 20, 18% by age of 30[7]. At age of 55 76% of all diagnosed patients were alive. Thromboembolism was the major cause of death in 80% of the deceased patients. Survival for pyridoxine responsive homocystinurics was significantly longer than for non-responsive patients, as, for instance, at age of 30 97% of the former versus 77% of the latter were alive. Moreover, as discussed in Chapter 1.6, pyridoxine responsive patients generally are suffering from more mild forms of the disease and present even untreated less symptoms of the syndrome or less severe abnormalities at later age than non-responsive patients do[7].

The evaluation of the effect of treatment with pyridoxine in about 300 pyridoxine responsive homocystinurics, as surveyed by Mudd[7], showed a statistically significant reduction of the number of initial and recurrent thromboembolic events. Moreover, the occurrence of ectopia lentis was lower in responsive patients after the start of pyridoxine therapy, although definitive evaluation of this effect awaits observation of more patients on treatment at early age[7]. In a small number of responsive patients a normal skeletal development occurred after early start of pyridoxine therapy[156]. The evaluation of the effect of pyridoxine on preventing and ameliorating osteoporosis awaits longitudinal studies of this condition in early treated patients[7]. Treatment from the first weeks of life might prevent mental damage as pyridoxine responsive patients detected by newborn screening and since then treated all showed average or above average intelligence[7]. Late onset treatment rarely reverses mental impairment, although in some cases it brought about behavioral improvement and slight increase of IQ[244].

The efficacy of early started methionine restriction and additional measures (Chapter 1.10.2, 1.10.3) in preventing mental retardation in pyridoxine non-responsive homocystinurics is clearly shown by Mudd's survey[7]. An improvement of about 35 points in mean IQ was observed in these patients if treated from the first weeks of life. Such treatment also reduced the incidence of seizures and the occurrence of ectopia lentis[7]. Assessment of the effect of early onset treatment in pyridoxine non-responders on thromboembolic events, osteoporosis, and mortality awaits longer follow-up in these patients. In late detected and therefore late treated pyridoxine non-responsive patients mental impairment did not change and seizures were not alleviated by methionine restriction[7]. Moreover, so far methionine restriction did not have reduced significantly the occurrence of thromboembolic events and ectopia lentis[7].

REFERENCES


1. Carson NAJ, Neill DW. Metabolic abnormalities detected in a survey of mentally backward individuals in Northern Ireland. Arch Dis Child 1962; 37: 505–13.
2. Field CMB, Carson NAJ, Cusworth DC, Dent CE, Neill DW. Homocystinuria: a new disorder of metabolism. Abstracts of the Tenth International Congress of Paediatricians (Lisbon), 1962, p. 274.
3. Gerritsen T, Vaughn JG, Waisman HA. The identification of homocystine in the urine. Biochem Biophys Res Commun 1962; 9: 493–6.
4. Carson NAJ, Cusworth DC, Dent CE, Field CMB, Neill DW, Westall RG. Homocystinuria: A new inborn error of metabolism associated with mental deficiency. Arch Dis Child 1963; 38: 425–36.
5. Gerritsen T, Waisman HA. Homocystinuria: an inborn error in the metabolism of methionine. Pediatrics 1964; 33: 413–20.
6. Mudd SH, Finkelstein JD, Irreverre F, Laster L. Homocystinuria: an enzymatic defect. Science 1964; 143: 1443–5.
7. Mudd SH, Skovby F, Levy HL, Pettigrew KD, Wilcken B, Pyeritz RE, Andria G, Boers GHJ, Bromberg IL, Cerone R, Fowler B, Gröbe H, Schmidt H, Schweitzer L. The natural history of homocystinuria due to cystathionine β-synthase deficiency. Am J Hum Genet 1985; 37: 1–31.
8. Carson NAJ. Homocystinuria: clinical and biochemical heterogeneity. In: Cockburn F, Gitzelmann R, eds. Inborn errors of metabolism in humans. Lancaster, England: MTP Press Limited, 1982: 53–67.
9. Wendel U, Bremer HJ. Betaine in the treatment of homocystinuria due to 5,10-methylenetetrahydrofolate reductase deficiency. Eur J Pediatr 1984; 142: 147–50.
10. Shinnar S, Singer HS. Cobalamin C mutation (methylmalonic aciduria and homocystinuria) in adolescence. N Engl J Med 1984; 311: 451–4.
11. Schuh S, Rosenblatt DS, Cooper BA, Schroeder ML, Bishop AJ, Seargeant LE, Haworth JC. Homocystinuria and megaloblastic anemia responsive to vitamin $B_{12}$ therapy. N Engl J Med 1984; 310: 686–90.
12. Linnell LC, Miranda B, Bhatt HR, Dowton SB, Levy HL. Abnormal cobalamin metabolism in a megaloblastic child with homocystinuria, cystathioninuria and methylmalonic aciduria. J Inher Metab Dis 1983; 6, Suppl 2: 137–9.
13. Ribes A, Vilaseca MA, Briones P, Maya A, Sabater J, Pascual P, Alvarez L, Ros J, Gonzalez Pascual E. Methylmalonic aciduria with homocystinuria. J Inher Metab Dis 1984; 7, Suppl. 2: 129–30.
14. Mudd SH, Poole JR. Labile methyl balances for normal humans on various dietary regimens. Metabolism 1975; 24: 721–35.
15. Finkelstein JD, Mudd SH. Trans-sulphuration in mammals: the methionine sparing effect of cystine. J Biol Chem 1967; 241: 873–80.
16. Finkelstein JD, Martin JJ. Methionine metabolism in mammals. Distribution of homocysteine between competing pathways. J Biol Chem 1984; 259: 9508–13.
17. Finkelstein JD, Harris BJ, Martin JJ, Kyle WE. Regulation of hepatic betaine-homocysteine methyltransferase by dietary methionine. Biochem Biophys Res Commun 1982; 108: 344–8.
18. Finkelstein JD, Martin JJ, Harris BJ, Kyle WE. Regulation of hepatic betaine-homocysteine methyltransferase by dietary betaine. J Nutr 1983; 113: 519–21.
19. Finkelstein JD, Kyle WE, Harris BJ. Methionine metabolism in mammals. Regulation of homocysteine methyltransferase in rat tissue. Arch Biochem Biophys 1971; 146: 84–92.
20. Racker E. Glutathione-homocystine transhydrogenase. J Biol Chem 1955; 217: 867–74.

21. Jocelyn PC. Biochemistry of the SH group. The occurrence, chemical properties, metabolism and biological function of thiols and disulphides. London, New York: Academic Press, 1972.
22. Brenton DP, Cusworth DC, Gaull GE. Homocystinuria: metabolic studies in 3 patients. J Pediatr 1965; 67: 58–68.
23. Perry TL, Hansen S, MacDougall L, Warrington PD. Sulphur-containing amino acids in the plasma and urine of homocystinurics. Clin Chim Acta 1967; 15: 409–20.
24. Kang SS, Wong PWK, Becker N. Protein-bound homocyst(e)ine in normal subjects and in patients with homocystinuria. Pediatr Res 1979; 13: 1141–43.
25. Kang SS, Wong PWK, Curley K. The effect of D-penicillamine on protein-bound homocyst(e)ine in homocystinurics. Pediatr Res 1982; 16: 370–2.
26. Werder EA, Curtius HCh, Trancredi F, Anders PW, Prader A. Homocystinurie. Helv Paediatr Acta 1966; 21: 1–18.
27. Boers GHJ, Smals AGH, Drayer JIM, Trijbels FJM, Leermakers AI, Kloppenborg PWC. Pyridoxine does not prevent homocystinemia after methionine loading in adult homocystinuria patients. Metabolism 1983; 32: 390–7.
28. Wilcken DEL, Gupta VJ. Cysteine-homocysteine mixed disulphide: differing plasma concentrations in normal men and women. Clin Sci 1979; 57: 211–5.
29. Boers GH, Smals AG, Trijbels FJ, Leermakers AI, Kloppenborg PW. Unique efficiency of methionine metabolism in premenopausal women may protect against vascular disease in the reproductive years. J Clin Invest 1983; 72: 1971–6.
30. Seashore MR, Durant JL, Rosenbergh LE. Studies of the mechanism of pyridoxine-responsive homocystinuria. Pediatr Res 1972; 6: 187–96.
31. Perry TL, Hansen S, Love DL, Crawford LE, Tischler B. Treatment of homocystinuria with a low-methionine diet, supplemental cystine, and a methyl donor. Lancet 1968; 2: 474–8.
32. Sardharwalla IB, Jackson SH, Hawke HD, Sass-Kortsak A. Homocystinuria: a study with low-methionine diet in three patients. Can Med Assoc J 1968; 99: 731–40.
33. Mudd SH, Levy HL. Disorders of transsulphuration. In: Stanbury JB, Wyngaarden JB, Fredrickson DS, eds. The metabolic basis of inherited disease. New York: McGraw-Hill Book Co ed. 4, 1978: 458–503.
34. Bremer HJ, Duran M, Kamerling JP, Przyrembel H, Wadman SK. Clinical chemistry and diagnosis of inherited diseases. In: Urban and Swarzenberg, eds. Disturbances of amino acid metabolism: clinical chemistry and diagnosis. Baltimore-München, 1981: 277–8.
35. Kennedy C, Shih VE, Rowland LP. Homocystinuria: a report in two siblings. Pediatrics 1965; 36: 736–41.
36. Spaeth GL, Barber GW. Homocystinuria in a mentally retarded child and her normal cousin. Trans Am Ophthalmol Oto-laryngol 1965; 69: 912–30.
37. Rassin DK, Longhi RC, Sternowsky HJ, Sturman JA, Gaull GE. Homocysteine and cysteine loads in patients with homocystinuria due to cystathionine synthase deficiency: effects of vitamin $B_6$. Clin Chim Acta 1977; 79: 197–210.
38. Mudd SH, Edwards WA, Loeb PM, Brown MS, Laster I. Homocystinuria due to cystathionine synthase deficiency: the effect of pyridoxine. J Clin Invest 1970; 49: 1762–73.
39. Barber GW, Spaeth GL. The successful treatment of homocystinuria with pyridoxine. J Pediatr 1969; 75: 463–78.
40. Laster L, Mudd SH, Finkelstein JD, Irreverre F. Homocystinuria due to cystathionine synthase deficiency: the metabolism of L-methionine. J Clin Invest 1965; 44: 1708–19.
41. Brenton DP, Cusworth DC. Homocystinuria: metabolism of ($^{35}S$) methionine. Clin Sci 1966; 31: 197–206.

42. Tada K, Yoshida T, Hirono H, Arakawa T. Homocystinuria: amino acid pattern of the liver. Tohoku J Exp Med 1967; 92: 325–32.
43. Tada K, Yoshida T, Arakada T. Free amino acid pattern in the liver from the patients with amino acid disorders: postmortem diagnosis of inborn errors of amino acid metabolism. Tohoku J Exp Med 1970; 101: 223–6.
44. Rassin DK, Longhi RC, Gaull GE. Free amino acids in liver of patients with homocystinuria due to cystathionine synthase deficiency: effects of vitamin $B_6$. J Pediatr 1977; 91: 574–7.
45. Mudd SH, Edwards WA. Unpublished data.
46. Tallan HH, Moore S, Stein WH. L-Cystathionine in human brain. J Biol Chem 1958; 230: 707–16.
47. Gerritsen T, Waisman HA. Homocystinuria: absence of cystathionine in the brain. Science 1964; 145: 588.
48. Brenton DP, Cusworth DC, Gaull GE. Homocystinuria: Biochemical studies of tissues including a comparison with cystathioninuria. Pediatrics 1965; 35: 50–56.
49. Hemraj F, Griffiths R. Enzyme studies in cystathionine β-synthase deficiency: a possible effect of elevated intracellular levels of homocystine in kinetic studies. J Inher Metab Dis 1978; 1: 171–4.
50. Shih VE, Mandell R, Levy HL, Littlefield JW. Free amino acids in extracts of cultured skin fibroblasts from patients with various amino acid metabolic disorders. Clin Genet 1975; 7: 421–5.
51. Vellan EJ, Gjessing LR, Seip M. Hair amino acids in cystinosis, homocystinuria, Fölling's disease and tyrosinosis. Acta Pediatr Scand 1969; 58: 287–9.
52. Kraus S, Packman S, Fowler B, Rosenberg LE. Purification and properties of cystathionine β-synthase from human liver. J Biol Chem 1978; 253: 6523–28.
53. Mudd SH, Finkelstein JD, Irreverre F, Laster L. Transsulphuration in mammals: microassays and tissue distributions of three enzymes of the pathway. J Biol Chem 1965; 240: 4382–92.
54. Uhlendorf BW, Mudd SH. Cystathionine synthase in tissue culture derived from human skin: enzyme defect in homocystinuria. Science 1968; 160: 1007–9.
55. Uhlendorf BW, Conerly EB, Mudd SH. Homocystinuria: studies in tissue culture. Pediatr Res 1973; 7: 645–8.
56. Goldstein JL, Campbell BK, Gartler SM. Cystathionine synthase activity in human lymphocytes: induction by phytohemagglutinin. J Clin Invest 1972; 51: 1034–7.
57. Fleisher LD, Beratis NG, Hirschhorn K, Gaull GE. Detection of cystathionine synthase in long-term lymphoid cell lines. Lancet 1972; 2: 482.
58. Skovby F, Kraus JP, Rosenberg LE. Biosynthesis and proteolytic activation of cystathionine β-synthase in rat liver. J Biol Chem 1984; 259: 588–93.
59. Skovby F, Kraus JP, Rosenberg LE. Biosynthesis of human cystathionine β-synthase in cultured fibroblasts. J Biol Chem 1984; 259: 583–7.
60. Skovby F, Krassikoff N, Franke U. Assignment of the gene for cystathionine β-synthase to human chromosome 21 in somatic cell hybrids. Hum Genet 1984; 65: 291–4.
61. Finkelstein JD, Mudd SH, Irreverre F. Laster L. Homocystinuria due to cystathionine synthase deficiency: the mode of inheritance. Science 1964; 146: 785–7.
62. Laster L, Spaeth GL, Mudd SH. Homocystinuria due to cystathionine synthase deficiency. Ann Intern Med 1965; 63: 1117–42.
63. Mudd SH, Finkelstein JD, Irreverre F, Laster L. Threonine dehydratase activity in humans lacking cystathionine synthase. Biochem Biophys Res Commun 1965; 19: 665–70.
64. Mudd SH, Laster L, Finkelstein JD. Irreverre F. Studies on homocystinuria. In: Himwich HE, Kety SS, Smythies JR, eds. Amines and schizophrenia. New York: Pergamon Press, 1965: 246–56.

65. Hollowell JG Jr. Coryell ME, Hall WK, Findley JK, Thevaos TG. Homocystinuria as affected by pyridoxine, folic acid, and vitamin $B_{12}$. Proc Soc Exper Biol Med 1968; 129: 327–33.
66. Yoshida T, Tada K, Yokoyama Y, Arakawa T. Homocystinuria of vitamin $B_6$ dependent type. Tohoku J Exp Med 1968; 96: 235–42.
67. Gaull GE, Rassin DK, Sturman JA. Enzymatic and metabolic studies of homocystinuria: effects of pyridoxine. Neuropaediatrie 1969; 1: 199–226.
68. Gaull GE, Sturman JA, Schaffner F. Homocystinuria due to cystathionine synthase deficiency: enzymatic and ultrastructural studies. J Pediatr 1974; 84: 381–90.
69. Longhi RC, Fleisher LD, Tallan HH, Gaull GE. Cystathionine β-synthase deficiency, qualitative abnormality of deficient enzyme modified by vitamin-$B_6$ therapy. Pediatr Res 1977; 11: 100–3.
70. Goldstein JL, Campbell BK, Gartler SM. Homocystinuria: heterozygote detection using phytohemagglutinin-stimulated lymphocytes. J Clin Invest 1973; 52: 218–21.
71. Fleisher LD, Beratis NG, Tallan HH, Hirschhorn K, Gaull GE. Homocystinuria due to cystathionine synthase deficiency: investigations in cultured long-term lymphocytes, fetal skin fibroblasts and amnion fluid cells. Pediatr Res 1974; 8: 388.
72. Bittles AH, Carson NAJ. Tissue culture techniques as an aid to prenatal diagnosis and genetic counselling in homocystinuria. J Med Genet 1973; 10: 120–1.
73. Porter PN, Grishaver MS, Jones OW. Characterization of human cystathionine β-synthase. Evidence for the identity of human L-serine dehydratase and cystathionine β-synthase. Biochim Biophys Acta 1974; 364: 128–39.
74. Poole JR, Mudd SH, Conerly EB, Edwards WA. Homocystinuria due to cystathionine synthase deficiency. Studies on nitrogen balance and sulphur excretion. J Clin Invest 1975; 55: 1033–48.
75. Fleisher LD, Longhi RC, Tallan HH, Gaull GE. Cystathionine β-synthase deficiency, differences in thermostability between normal and abnormal enzyme from cultured human cells. Pediatr Res 1978; 12: 293–6.
76. Fowler B, Kraus J, Packman S, Rosenberg LE. Homocystinuria. Evidence of 3 distinct classes of cystathionine β-synthase mutants in cultured fibroblasts. J Clin Invest 1978; 61: 645–53.
77. Drayer JIM, Cleophas AJM, Trijbels JMF, Smals AGH, Kloppenborg PWC. Symptoms, diagnostic pitfalls, and treatment of homocystinuria in seven adult patients. Neth J Med 1980; 23: 89–94.
78. Bittles AH, Carson NAJ. Homocystinuria: studies on cystathionine β-synthase, S-adenosylmethionine synthetase and cystathionase activities in skin fibroblasts. J Inher Metab Dis 1981; 4: 3–6.
79. Boers GHJ, Smals AGH, Trijbels JMF, Schoonderwaldt HC, Bakkeren JAJM, Kleyer WJ, Fowler B, Kloppenborg PWC. High frequency of heterozygosity for homocystinuria in premature occlusive peripheral and cerebral vascular disease. Submitted for publication.
80. Barber GW, Spaeth GL. Pyridoxine therapy in homocystinuria. Lancet 1967; 1: 337.
81. Hooft C, Carton D, Samyn W. Pyridoxine treatment in homocystinuria. Lancet 1967; 1: 1384.
82. Carson NA, Carré IJ. Treatment of homocystinuria with pyridoxine. Arch Dis Child 1969; 44: 387–92.
83. Mudd SH, Levy HL. Disorders of transsulpuration. In: Stanbury JB, Wyngaarden JB, Fredrickson DS, eds. The metabolic basis of inherited disease. New York: McGraw-Hill Book Co, ed. 5, 1983: 522–59.
84. Kim Y, Rosenberg LE. On the mechanism of pyridoxine responsive homocystinuria. II. Properties of normal and mutant cystathionine β-synthase from cultured fibroblasts. Proc Natl Acad Sci U.S.A. 1974; 71: 4821–5.

85. Skovby F, Kraus J, Redlich C, Rosenberg LE. Immunochemical studies on cultured fibroblasts from patients with homocystinuria due to cystathionine β-synthase deficiency. Am J Hum Genet 1982; 34: 73–83.
86. Skovby F, Kraus JP, Rosenberg LE. Homocystinuria: Biogenesis of cystathionine β-synthase subunits in cultured fibroblasts and in an in vitro translation system programmed with fibroblast messenger RNA. Am J Hum Genet 1984; 36: 452–9.
87. Fowler B, Sardharwalla IB. Homocystinuria: cystathionine synthase activity in cultured skin fibroblasts: In: International Symposium on inborn errors of metabolism in humans, Switzerland 1980: 20.
88. Lipson MH, Kraus J, Rosenberg LE. Affinity of cystathionine β-synthase for pyridoxal 5'phosphate in cultured cells. J Clin Invest 1980; 66: 188–93.
89. Hollowell JG Jr, Hall WK, Coryell ME, McPherson J Jr, Hahn DA. Homocystinuria and organic aciduria in a patient with vitamin-$B_{12}$ deficiency. Lancet 1969; 2: 1428.
90. Mudd SH. Homocystinuria and homocysteine metabolism: selected aspects. In: Nyan WL, ed. Heritable disorders of amino acid metabolism. New York: John Wiley and Sons, Inc, 1974: 429–51.
91. Higginbottom MC, Sweetman L, Nyan WL. A syndrome with methylmalonuric aciduria in a vitamin $B_{12}$-deficient breast-fed infant of a strict vegetarian. N Engl J Med 1978; 299: 317–23.
92. Levy HL, Mudd SH. Homocystinuria due to bacterial contamination in pyridoxine unresponsive cystathioninemia. Pediatr Res 1973; 7: 390.
93. Levy HL, Mudd SH, Uhlendorf BW, Madigan PM. Cystathioninuria and homocystinuria. Clin Chim Acta 1975; 58: 51–9.
94. Lyon ICT, Procopis PG, Turner B. Cystathioninuria in a well baby population. Acta Paediatr Scand 1971; 60: 324–8.
95. Przyrembel H, Bremer HJ. Cystathioninuria in premature infants. Clin Chim Acta 1972; 41: 95–9.
96. Endres W, Seibold H. Renal excretion of cystathionine and creatinine in humans at different ages. Clin Chim Acta 1978; 87: 425–32.
97. Endres W, Vogt R, Riegel KP, Bremer HG. The influence of vitamin $B_6$ on cystathioninuria in premature infants. Clin Chim Acta 1978; 86: 89–93.
98. Scriver CR, Hutchinson JH. The vitamin $B_6$ deficiency syndrome in human infancy: biochemical and clinical observations. Pediatrics 1963; 31: 240–50.
99. Sturman JA, Cohen PA, Gaull GE. Effects of deficiency of vitamin $B_6$ on transsulfuration. Biochem Med 1969; 3: 244–51.
100. Park YK, Linkswiler H. Effect of vitamin $B_6$ depletion in adult man on the excretion of cystathionine and other methionine metabolites. J Nutr 1970; 100: 110–6.
101. Finkelstein JD, Chalmers FT. Pyridoxine effects on cystathionine in rat liver. J Nutr 1970; 100: 467–9.
102. Shaw KNF, Lieberman E, Koch R, Donnell GN. Cystathioninuria. Am J Dis Child 1967; 113: 119–28.
103. Lieberman E, Shaw KNF, Donnell GN. Cystathioninuria in galactosemia, and certain types of liver disease. Pediatrics 1967; 40: 828–33.
104. Von Studnitz W. Secondary cystathioninuria. Acta Paediatr Scand 1969; 58: 173–7.
105. Endres W, Wuttage B. Occurrence of secondary cystathioninuria in children with inherited metabolic disorders, liver diseases, neoplasms, cystic fibrosis and celiac disease. Eur J Pediatr 1978; 129: 29–35.
106. Gjessing LR. Cystathioninuria during a load of thyroxine. Scand J Clin Lab Invest 1964; 16: 680–1.
107 Gjessing LR. Studies of functional neural tumors. II. Cystathioninuria. Scand J Clin Lab Invest 1963; 15: 474–8.

108. Gjessing LR. Studies on functional neural tumors. IV. Isolation and identification of urinary cystathionine. Scand J Clin Lab Invest 1963; 15: 601–2.
109. Gjessing LR, Mauritzen K. Cystathioninuria in hepatoblastoma. Scand J Clin Lab Invest 1965; 17: 513–4.
110. Voûte PA Jr, Wadman SK. Cystathioninuria in hepatoblastoma. Clin Chim Acta 1968; 22: 373–8.
111. Geiser CF, Shih VE. Cystathioninuria and its origin in children with hepatoblastoma. J Pediatr 1980; 96: 72–5.
112. Guthrie R. Screening for "inborn errors of metabolism" in the newborn infant, a multiple test program. Birth Defects Original Article Series 1968; 4: 92–8.
113. Levy HL, Shih VE, MacCready RA. Screening for homocystinuria in the newborn and mentally retarded population. In: Carson NAJ, Raine DN, eds. Inherited disorders of sulphur metabolism. Edinburgh: Churchill Livingstone, 1971: 235–43.
114. Whiteman PD, Clayton BE, Ersser RS, Lilly P, Seakins JWT. Changing incidence of neonatal hypermethioninaemia: implications for the detection of homocystinuria. Arch Dis Child 1979; 54: 593–8.
115. Sardharwalla IB. Comment in Whiteman PD, Clayton BE, Ersser RS, Lilly P, Seakins JWT: Changing incidence of neonatal hypermethioninaemia: implications for the detection of homocystinuria. Arch Dis Child 1979; 54: 598.
116. Brand E, Harris MM, Biloon S. Cystinuria. The excretion of cystine complex with decomposes in the urine with the liberation of free cystine. J Biol Chem 1930; 86: 315–31.
117. McKusick VA. Homocystinuria. In: McKusick VA, ed. Heritable disorders of connective tissue. St. Louis: CV Mosby Co, ed. 4, 1972: 224–81.
118. Mudd SH. Discussion. In: Carson NAJ, Raine DN, eds. Inherited disorders of sulphur metabolism. London: Churchill Livingstone Ltd, 1971: 311.
119. Fleisher LD, Longhi RC, Tallan HH, Beratis NG, Hirschhorn K, Gaull GE. Homocystinuria: investigations of cystathionine synthase in cultured fetal cells and the prenatal determination of genetic status. J Pediatr 1974; 85: 677–80.
120. Kurczynski TW, Muir WA, Fleisher LD, Palomaki JF, Gaull GE, Rassin DK, Abramowsky C. Maternal homocystinuria: studies of an untreated mother and fetus. Arch Dis Child 1980; 55: 721–3.
121. Fowler B, Børresen AL, Boman N. Prenatal diagnosis of homocystinuria. Lancet 1982; 2: 875.
122. Dunn HG, Perry ThL, Dolman CL. Homocystinuria. A recently discovered cause of mental defect and cerebrovascular thrombosis. Neurology 1966; 16: 407–20.
123. Wilcken B, Turner G. Homocystinuria in New South Wales. Arch Dis Child 1978; 53: 242–5.
124. McKusick VA, Hall JG, Char F. The clinical and genetic characteristics of homocystinuria. In: Carson NAJ, Raine DN, eds. Inherited disorders of sulphur metabolism. London: Churchill Livingstone Ltd., 1971: 179–203.
125. Boers GHJ, Polder TW, Cruysberg JRM, Schoonderwaldt HC, Peetoom JJ, van Ruyven TWJ, Smals AGH, Kloppenborg PWC. Homocystinuria versus Marfan's syndrome: the therapeutic relevance of the differential diagnosis. Neth J Med 1984; 27: 206–12.
126. Turner G, Dey J, Turner B. Homocystinuria: A report of two Australian families. Aust Pediatr J 1967; 3: 48–53.
127. Wicherink-Bol HF, Boers GHJ, Drayer JIM, Rosenbursch G. Angiographic findings in homocystinuria. Cardiovasc Intervent Radiol 1983; 6: 125–8.
128. Ramsey MS, Yanoff M, Fine BS. The ocular histopathology of homocystinuria. A light and electron microscopic study. Am J Ophthalmol 1972; 74: 377–85.
129. Ramsey MS, Daitz LD, Beaton JW. Lens fringe in homocystinuria. Arch Ophthalmol 1975; 93: 318.

130. Francois J. Ocular manifestations in aminoacidopathies. Adv Ophthalmol 1972; 25: 28–108.
131. Cross HE, Jensen AD. Ocular manifestations in the Marfan syndrome and homocystinuria. Am J Ophthalmol 1973; 75: 405–20.
132. Maumenee IH. The eye in the Marfan syndrome. Trans Am Ophthalmol Soc 1981; 79: 684–733.
133. Lieberman TW, Podos SM, Hartstein J. Acute glaucoma, ectopia lentis and homocystinuria. Am J Ophthalmol 1966; 61: 252–5.
134. Thomas RP, Hollowell JG, Peters HJ, Coryell ME, Lester RH. Homocystinuria and ectopia lentis in negro family. JAMA 1966; 198: 560–2.
135. Elkington AR, Freedman SS, Jay B, Wright P. Anterior dislocation of the lens in homocystinuria. Br J Ophthalmol 1973; 57: 325–9.
136. Schoonderwaldt HC, Boers GHJ, Cruysberg JRM, Schulte BPM, Slooff JL, Thijssen HOM. Neurologic manifestations of homocystinuria. Clin Neurol Neurosurg 1981; 83: 153–62.
137. Mukuno K, Matsui K, Haraguchi H. Ocular manifestations of homocystinuria: report of two cases. Acta Soc Ophthalmol Jap 1967; 71: 66–73.
138. Wilson RS, Ruiz RS. Bilateral central retinal artery occlusion in homocystinuria. Arch Ophthalmol 1969; 82: 267–8.
139. Spaeth GL, Barber GW. Homocystinuria, its ocular manifestations. J Pediatr Ophthalmol 1966; 3: 42–8.
140. Presley GD, Sidbury JB. Homocystinuria and ocular defects. Am J Ophthalmol 1967; 63: 1723–7.
141. Smith TH, Holland MG, Woody NC. Ocular manifestations of familial hyperlysinemia. Trans Am Acad Ophthalmol Oto-laryngol 1971; 75: 355–60.
142. Irreverre F. Mudd SH, Heizer WD, Laster L. Studies of a patient with mental retardation, dislocated ocular lenses and abnormal urinary excretion of S-sulfo-L-cysteine, sulfite and thiosulfate. Biochem Med 1967; 1: 187–217.
143. Shih VE, Abrams IF, Johnson IL, Mandell R, Robb RM, Cloherty IP, Rajagopalan KV. Sulfite oxidase deficiency. Biochemical and clinical investigations of a hereditary metabolic disorder in sulfur metabolism. N Engl J Med 1977; 297: 1022–8.
144. Beemer FA, Delleman JW. Combined deficiency of xanthine oxidase and sulfite oxidase: ophthalmological findings in a 3-week old girl. Metab Pediatr Ophthalmol 1980; 4: 49–52.
145. Nelson LB, Maumenee IH. Ectopia lentis. Survey Ophthalmol 1982; 27: 143–60.
146. Smith SW. Roentgen findings in homocystinuria. Am J Roentgenol Radium Ther Nucl Med 1967; 100: 147–54.
147. Schedewie H, Willich E, Gröbe H, Schmidt H, Müller KM. Skeletal findings in homocystinuria: a collaborative study. Pediatr Radiol 1973; 1: 12–23.
148. Morreels CL, Fletcher BD, Weilbaecher RG, Dorst JP. The roentgenographic features of homocystinuria. Radiology 1968; 90: 1150–8.
149. Gröbe H, v. Bassewitz DB. Thromboembolische Komplikationen und Thrombozytenanomalien bei Homocystinurie. Z. Kinderheilk 1972; 112: 309–20.
150. Brenton DP, Dow CJ, James JIP, Hay RL, Wynne-Davies R. Homocystinuria and Marfan's syndrome, a comparison. J Bone Joint Surg 1972; 54: 277–98.
151. Beals RK. Homocystinuria. A report of two cases and review of the literature. J Bone Joint Surg 1969; 51: 1564–72.
152. Schimke NR, McKusick VA, Huang Th, Pollack AD. Homocystinuria. A study of 20 families with 38 affected members. JAMA 1965; 193: 711–9.
153. Gaudier B, Remy J, Nuyts JP, Caron-Poitreau C, Bombart E, Foissac-Gegoux MC. Etude radiologique des signes osseux de l'homocystinurie. Arch Fr Pediatr 1969; 26: 963–75.

154. Brenton DP, Cusworth DC. The response of patients with cystathionine synthase deficiency to pyridoxine. In: Carson NAJ, Raine DN, eds. Inherited disorders of sulphur metabolism. Edinburgh: Churchill Livingstone, 1971: 264–74.
155. Brenton DP. Skeletal abnormalities in homocystinuria. Post Grad Med J 1977; 53: 488–94.
156. Thomas PS, Carson NAJ. Homocystinuria. The evolution of skeletal changes in relation to treatment. Ann Radiol 1978; 21: 95–104.
157. Carson NAJ, Dent DE, Field CMB, Gaull GE. Homocystinuria. Clinical and pathological review of 10 cases. J Pediatr 1965; 66: 565–83.
158. Gaull GE, Sturman JA. Vitamin $B_6$ dependency in homocystinuria. Br Med J 1971; 3: 532–3.
159. Hopkins J, Townley RRW, Shipman RT. Cerebral thrombosis in a patient with homocystinuria. J Pediatr 1969; 75: 1082–84.
160. Carey MC, Donovan DE, Fitzgerald O, McAuley FD. Homocystinuria. I. A clinical and pathological study of nine subjects in six families. Am J Med 1968; 45: 7–25.
161. Cusworth DC, Dent CE. Homocystinuria. Br Med Bull 1969; 25: 42–7.
162. Gibson JB, Carson NAJ, Neill DW. Pathological findings in homocystinuria. J Clin Pathol 1964; 17: 427–37.
163. Komrower GM, Wilson VK. Homocystinuria. Proc R Soc Med 1963; 56: 996–7.
164. Brenton DP, Cusworth DC, Dent CE, Jones EE. Homocystinuria, clinical and dietary studies. Q J Med 1966; 35: 325–46.
165. Brill PW, Mitty HA, Gaull GE. Homocystinuria due to cystathionine synthase deficiency: clinical-roentgenologic correlations. Am J Roentgenol Radium Ther Nucl Med 1974; 121: 45–54.
166. McCully KS. Vascular pathology of homocysteinemia: implications for the pathogenesis of arteriosclerosis. Am J Pathol 1969; 56: 111–28.
167. Almgren B, Eriksson I, Hemmingsson A, Hillerdal G, Larsson E, Âberg H. Abdominal aortic aneurysm in homocystinuria. Acta Chir Scand 1978; 144: 545–8.
168. Newman G, Mitchell JRA. Homocystinuria presenting as multiple arterial occlusions. Q J Med 1984; 210: 251–8.
169. Brown RB, Watson PD, Taussig LM. Congenital metabolic diseases of pediatric patients: anesthesiological implications. Anesthesiology 1975; 43: 197–209.
170. Regenbogen L, Ilie S, Elian I. Homocystinuria, a surgical and anesthetic risk. Metab Pediatr Ophthalmol 1980; 4: 209–12.
171. Crooke JW, Towers JF, Taylor WH. Management of patients with homocystinuria requiring surgery under general anaesthesia. Br J Anaesth 1971; 43: 96–9.
172. Parris WCV, Quimby CW. Anesthetic considerations for the patient with homocystinuria. Anesth Analg (Cleve) 1982; 61: 708–10.
173. Jackson GM, Grisolia JS, Wolf PL, Jones OW, Bloor CM. Postoperative thromboemboli in cystathionine β-synthase deficiency. Am Heart J 1984; 108: 627–8.
174. James ThN, Carson NAJ, Froggaat P. The subitaneis mortibus. IV. Coronary vessels and conduction system in homocystinuria. Circulation 1974; 49: 367–74.
175. Shih VE, Efron ML. Pyridoxine-unresponsive homocystinuria. N Engl J Med 1970; 283: 1206–9.
176. White HH, Rowland LP, Araki S, Thompson HL, Cowen D. Homocystinuria. Arch Neurol 1965; 13: 455–70.
177. Hagberg B, Hambraeus L, Bensch K. A case of homocystinuria with a dystonic neurological syndrome. Neuropaediatrie 1970; 1: 337–43.
178. Davous P, Rondot P. Homocystinuria and dystonia. J Neurol 1983; 46: 283.
179. Gale AN, Brenton DP, Cusworth DC, Duchen LW, Stern GM. Extrapyramidal disturbance in homocystinuria. J Med Genet 1982; 19: 374.

180. Gröbe H. Homocystinurie. Klinisches Bild, Behandlung und Ergebnisse bei 8 Patienten. Dtch Med Wochenschr 1973; 98: 1313–9.
181. Komrower GM, Sardharwalla IB. The dietary treatment of homocystinuria. In: Carson NAJ, Raine DN, eds. Inherited disorders of sulphur metabolism. Edinburgh: Churchill-Livingstone, 1971: 254–63.
182. Gaull GE, Schaffner F. Electron microscopic changes in hepatocytes of patients with homocystinuria. Pediatr Res 1971; 5: 23–32.
183. McCully KS. Importance of homocysteine-induced abnormalities of proteoglycan structure in arteriosclerosis. Am J Path 1970; 59: 181–93.
184. Jackson SH. The reaction of homocysteine with aldehyde: an explanation of the collagen defects in homocystinuria. Clin Chim Acta 1973; 45: 215–7.
185. Strydom AJC, Meiring JL. A mechanism for the tissue damage in homocystinuria. In: Abstracts of the Third International Symposium on inborn errors of metabolism in humans. Munich: Karger, 1984: 81.
186. Harris ED, Sjoerdsma A. Collagen profile in various clinical conditions. Lancet 1966; 2: 707–11.
187. Kang AH, Trelstad RL. A collagen defect in homocystinuria. J Clin Invest 1973; 52: 2571–8.
188. Streeten BW, Swann DA, Licari PA, Robinson MR, Gibson SA, Marsh NJ, Vergnes JP, Freeman IL. The protein composition of the ocular zonules. Invest Ophthalmol Vis Sci 1983; 24: 119–23.
189. Arnott EJ, Greaves DP. Ocular involvement in homocystinuria. Br J Ophthalmol 1964; 48: 688–9.
190. McDonald L, Bray C, Field C, Love F, Davies B, Homocystinuria, thrombosis, and the blood platelets. Lancet 1964; 1: 745–6.
191. Brett EM. Homocystinuria with epilepsy. Proc R Soc Med 1966; 59: 484–7.
192. Chase HP, Goodman SI, O'Brien D. Treatment of homocystinuria. Arch Dis Child 1967; 42: 514–20.
193. Holmgren G, Hambraeus L, Lestrup E, Tangen O. The effect of pyridoxine on platelet adhesiveness in homocystinuria. Neuropaediatrie 1974; 5: 402–9.
194. Harker LA, Slichter SJ, Scott CR, Ross R. Homocystinemia. Vascular injury and arterial thrombosis. N Engl J Med 1974; 291: 537–43.
195. Hampton JR, Mitchel JRA. A transferable factor causing abnormal platelet behaviour in vascular disease. Lancet 1966; 2: 764–8.
196. Zweifler AJ, Allen RJ. An intrinsic blood platelet abnormality in an homocystinuric boy, corrected with pyridoxine administration. Thromb Diath Haemorrh 1971; 26: 15–21.
197. Davis JW, Lawrence D, Flournoy D, Phillips PhE. Amino acids and collagen-induced platelet aggregation. Am J Dis Child 1975; 129: 1020–1.
198. Uhlemann ER, TenPas JH, Lucky AW, Schulman JD, Mudd SH, Shulman NR. Platelet survival and morphology in homocystinuria due to cystathionine synthase deficiency. N Engl J Med 1976; 295: 1283–6.
199. Ratnoff OD. Activation of Hageman factor by L-homocystine. Science 1968; 162: 1007–9.
200. Palareti G, Salardi S, Tricarico MG, Poggi M, Grauso F, Piazzi S, Cacciari E, Coccheri A. Antithrombin III deficiency in homocystinuria and its reversal during therapeutic correction of the metabolic defect. Hemostatis 1982; 12: 95.
201. Merckx J, Kuntz F. Déficit en facteur VII et homocystinurie. Nouv Press Med 1981; 46: 3796.
202. Munnich A, Saudubray JM, Dautzenberg MD, Parvy Ph, Ogier H, Girot R, Manigne P, Frézal J. Diet-responsive proconvertin (factor VII) deficiency in homocystinuria. J Pediatr 1983; 102: 730–4.

203. Gaull GE. Homocystinuria. Adv Teratol 1967; 2: 101–26.
204. McCully KS, Ragsdale BD. Production of arteriosclerosis by homocysteinemia. Am J Path 1970; 61: 1–8.
205. McCully KS, Wilson RB. Homocysteine theory of arteriosclerosis. Atherosclerosis 1975; 22: 215–27.
206. Harker LA, Ross R, Slichter SJ, Scott CR. Homocystine-induced arteriosclerosis. The role of endothelial cell injury and platelet response in its genesis. J Clin Invest 1976; 58: 731–41.
207. Harker LA, Ross R. Prevention of homocysteine induced arteriosclerosis: sulphinpyrazone endothelial protection. In: Abe T, Sherry S, eds. A new approach to reduction of cardiac death. Bern-Stuttgart-Vienna: Hans Huber Publishers, 1979: 59–72.
208. Harker LA, Harlan JM, Ross R. Effect of sulfinpyrazone on homocysteine-induced endothelial injury and arteriosclerosis in baboons. Circ Res 1983; 53: 731–9.
209. Hill-Zobel RL, Pyeritz RE, Schefel J, Malpica O, Engin S, Camargo EE, Abbot M, Guilarte TR, Hill J, McIntyre PA, Murphy EA, Tsan MF. Kinetics and distribution of $^{111}$Indium-labeled platelets in patients with homocystinuria. N Engl J Med 1982; 307: 781–6.
210. Harker LA, Scott CR. Platelets in homocystinuria. N Engl J Med 1977; 296: 818.
211. Terebelo HR. $^{111}$Indium-labeled platelets in homocystinuria. N Engl J Med 1983; 308: 4–5.
212. Wall RT, Harlan JM, Harker LA, Striker GE. Homocysteine-induced endothelial cell injury in vitro: a model for the study of vascular injury. Thromb Res 1980; 18: 113–21.
213. De Groot PhG, Willems Ch, Boers GHJ, Gonsalves MD, van Aken WG, van Mourik JA. Endothelial dysfunction in homocystinuria. Eur J Clin Invest 1983; 13: 405–10.
214. Harker LA, Ross R. Sulphydryl-mediated vascular disease. Eur J Clin Invest 1978; 8: 199.
215. Curtis DR, Watkins JC. Acidic amino acids with strong excitatory actions on mammalian neurons. J Physiol (London) 1963; 166: 1–14.
216. Gaitonde JK. Sulphur amino acids. In: Latha A, ed. Handbook of Neurochemistry, vol. III. New York: Plenum Press, 1970: 255–87.
217. Sprince H, Parker CM, Josephs JH. Homocysteine-induced convulsions in the rat: protection by homoserine, serine, betaine, glycine and glucose. Agents Act 1969; 1: 9–13.
218. Sprince H, Parker CM, Josephs JH, Magazino J. Convulsant activity of homocysteine and other short-chain mercapto acids: protection therefrom. Ann NY Acis Sci 1969; 166: 323–5.
219. Folbergrova J. Energy metabolism in mouse cerebral cortex during homocysteine convulsions. Brain Res 1974; 81: 443–54.
220. Blennow G, Folbergrova J, Nilsson G, Siesjo BK. Cerebral metabolic and circulatory changes in the rat during sustained seizures induced by DL-homocysteine-convulsions. Brain Res 1979; 179: 129–46.
221. Hammond EJ, Hurd RW, Wilder BJ, Thompson FJ. Focal and generalized experimental seizures induced by homocysteine. Electroenceph Clin Neurophys 1980; 49: 184–6.
222. Hurd RW, Hammond EJ, Wilder BJ. Homocysteine induced convulsions: enhancement by vitamin $B_6$ and inhibition by hydrazine. Brain Res 1981; 209: 250–4.
223. Dewhurst IC, Hagan JJ, Morris RGM, Griffiths R. Hippocampal electrical activity and γ.aminobutyrate metabolism in brain tissue following administration of homocysteine. J Neurochem 1983; 40: 752–7.
224. Griffiths R, Williams DC, O'Neill C, Dewhurst IC, Ekuwem CE, Sinclair CD. Synergistic inhibition of ($^3$H) muscimol binding to calf-brain synaptic membranes in the presence of L-homocysteine and pyridoxal 5'phosphate. Eur J Biochem 1983; 137: 467–78.
225. Thalhammer O. Frequency of inborn errors of metabolism in some representative

newborn screening centres around the world: a collaborative study. Hum Genet 1975; 30: 273–86.
226. Komrower GM, Sardharwalla IB, Fowler B, Bridge C. The Manchester regional screening program: a 10-year excercise in patient and family care. Br Med J 1979; 2: 635–8.
227. Johnston SS. Homocystinuria. Ophthalmologica 1978; 176: 282–3.
228. Spaeth GL, Barber GW. Prevalence of homocystinuria among the mentally retarded: evaluation of a specific screening test. Pediatrics 1967; 40: 586–9.
229. Komrower GM, Lambert AM, Cusworth DC, Westfall RG. Dietary treatment of homocystinuria. Arch Dis Child 1966; 41: 666–71.
230. Perry TL, Dunn HG, Hansen S, MacDougall L, Warrington PD. Early diagnosis and treatment of homocystinuria. Pediatrics 1966; 37: 502–5.
231. Byrd VJ, Heinbokel H, Schweitzer S, Brodehl J. Evaluation of oral L-cysteine for treatment of cystine deficiency in homocystinuria. In: Abstracts of the Third International Symposium on inborn errors of metabolism in humans. Munich: Karger, 1984: 104.
232. Parkinson MS. Therapeutic problems of adolescent homocystinuria. Proc R Soc Med 1969; 62: 909–10.
233. Carey MC, Fennelly JJ, Fitzgerald O. Homocystinuria. II. Subnormal serum folate levels, increased folate clearance and effects of folic acid therapy. Am J Med 1968; 45: 26–31.
234. Morrow G III, Barness LA. Combined vitamin responsiveness in homocystinuria. J Pediatr 1972; 81: 946–54.
235. Smolin LA, Benevenga NJ, Berlow S. The use of betaine in the treatment of homocystinuria. J Pediatr 1981; 99: 467–72.
236. Wilcken DEL, Wilcken B, Dudman NPB, Tyrrell PA. Homocystinuria. The effect of betaine in the treatment of patients not responsive to pyridoxine. N Engl J Med 1983; 309: 448–53.
237. Benevenga NJ. Betaine in the treatment of homocystinuria. N Engl J Med 1984; 310: 265–6.
238. Schulman JD, Agarwal B, Mudd SH, Shulman NR. Pulmonary embolism in a homocystinuric patient during treatment with dipyridamole and acetylsalicylic acid. N Engl J Med 1978; 299: 661.
239. Schulman JD, Mudd SH, Shulman NR, Landvater L. Pregnancy and thrombophlebitis in homocystinuria. Blood 1980; 56: 326.
240. Perry TL. Homocystinuria. In: Nyan WL, ed. Heritable disorders of amino acid metabolism. New York: John Wiley & Sons, Inc, 1974: 395–428.
241. Brenton DP, Cusworth DC, Biddle SA, Garrod PJ, Lasley L. Pregnancy and homocystinuria. Ann Clin Biochem 1977; 14: 161–2.
242. Gröbe H. Homocystinuria and oral contraceptives. Lancet 1978; 1: 158–9.
243. McCully KS. Homocystine, atherosclerosis and thrombosis: implications for oral contraceptive users. Am J Clin Nutr 1975; 28: 542–9.
244. Gröbe H. Homocystinuria (cystathionine synthase deficiency). Results of treatment in late-diagnosed patients. Eur J Pediatr 1980; 135: 199–203.

*Chapter 2*

# Clinical and biochemical findings in 21 homozygous homocystinuria patients

## 2.1 INTRODUCTION

Since the earlier described morbid history of the first adult patient with homozygous homocystinuria diagnosed in 1977[1], 16 additional patients aging between 9 and 54 years were detected until the end of 1984. Two homocystinuric brothers, patients nos 10 and 11, had been diagnosed at ages 4 and 3 years at the department of Pediatrics of this hospital in 1970[2]. They were treated since then and were referred to our department in 1982 for further follow-up when they had reached adult age. In addition, two other homocystinurics, patients nos 8 and 21, have been referred to our clinic in 1980 and 1984 after establishment of the diagnosis elsewhere (in patient no 8 by Drs. K.S. Liem, St. Jans Gasthuis, Hoorn, in 1968, and in patient no 21 by Dr. M.L.M. Houben, de Wever Ziekenhuis, Heerlen, in 1975). They were treated since diagnosis and referred to our department for further evaluation, bringing up the total number of diagnosed or treated homocystinuria patients to 21. Twelve patients were females, 9 were men. The 21 patients originated from 16 sibships. Data of the patients nos 1 to 13 have been included in the recent collaborative study by Mudd et al[3] who surveyed the natural histories of more than 600 homocystinuria patients diagnosed in the U.S.A., Canada, Australia, and in a number of European countries.

Nineteen of the 21 homocystinuria patients reside at the area surrounding this hospital. Conclusions on the incidence of homocystinuria in this region, however, based upon this number of diagnosed patients are not reliable because in the Netherlands newborn screening, although conducted for more than 10 years, does not include screening for homocystinuria.

2.2 DELAY BETWEEN FIRST SIGNS AND DEFINITIVE DIAGNOSIS OF HOMOCYSTINURIA (Table 1)

In 20 out of the 21 patients abnormalities which could have been suggestive of the diagnosis homocystinuria had been established at ages ranging from 3 months to 40 years (Table 1). In 13 patients ectopia lentis or its complication acute pupillary block glaucoma had been diagnosed at ages 1 to 36 years. Nevertheless, in 19 of these 20 patients, even in those with ectopia lentis, the definitive diagnosis of homocystinuria was delayed for periods ranging from 2 to 32 years. Major reasons for the long delay between first signs and diagnosis were: i. ignorance of the rather high frequency of homocystinuria in patients with ectopia lentis as our group reported earlier[4], ii. resemblance of the clinical picture of homocystinuria to that in Marfan's syndrome, leading to incorrect diagnosis in 40% of our patients, iii. inadequate biochemical screening for homocystinuria mostly based on the so called Brand test (Chapter 1.5 and 2.3) and, iv. the rather variegated presented pathology, ranging from severe psychiatric disorders to vascular events at young age such as thrombophlebitis, pulmonary embolism, and carotid artery stenosis (Table 1).

In 15 out of the 21 patients the diagnosis homocystinuria was made in adults aged 20 to 54 years. It is of note that all these patients proved to be pyridoxine responsive (Chapter 2.3) whereas the patients diagnosed before adultness did not or did only partially respond to pyridoxine treatment. The late detection of pyridoxine responsive homocystinurics is most probably due to the observation that responders generally present milder forms of the disease with fewer and less severe symptoms occurring at older age than non-responders (Chapter 1.6). Therefore, it is necessary that the possibility of homozygous homocystinuria should be considered also in adults with only minor or few signs of the multisystem disorder.

The conditions leading to investigations for the presence of homocystinuria in our 21 patients were: i. ectopia lentis as the only feature in 11 patients (52%) and as contributory symptom in 3 further patients (14%), ii. vascular events at young age in 2 patients (10%), iii. marfanoid features, psychomotor retardation, and psychotic depression in each one patient (5%), and, iv. established homocystinuria in a sib in 5 patients (24%). These conditions and the percentages of patients in whom they were the sole or major cause leading to investigation for homocystinuria, are about similar to those established by Mudd et al[3] in his survey on more than 600 homocystinurics.

2.3 BIOCHEMICAL FINDINGS (Table 2)

Table 2 shows the fasting levels of total homocysteine, methionine and cystine in the 21 homozygous homocystinuria patients. In all but two,

patients nos 12 and 15, the typical biochemical features of homozygosity for homocystinuria due to cystathionine synthase deficiency were present: elevated levels of total homocysteine up to 274 μmol/l, increased methionine concentrations as high as 549 μmol/l and decreased cystine levels as low as non-detectable in 2 patients.
In patient no 8, a man presenting marked clinical symptoms, only slight homocysteinemia with normal methionine and low-normal cystine levels was established in the fasting state. However, the rise of total homocysteine after methionine loading as well as the residual cystathionine synthase activity in cultured fibroblasts in this patient undoubtedly disclosed his homozygous state (Table 3).
It is of note that patient no 9, a 20-year old female with severe and fatal complications of homocystinuria[6], showed only a moderate hypermethioninemia in the presence of extremely high homocysteinemia and very low cystine levels. Fasting for several days due to her illness before sampling blood, might have accounted for this finding.
Patients nos 12 and 15 presented a normal profile of amino acid levels in the fasting state. The diagnosis of homozygous homocystinuria in these patients was based upon the abnormal rises of total homocysteine after methionine loading and their residual cystathionine synthase activities (Table 3), as will be discussed further in Chapter 2.4.

The qualitative and still popular cyanide-nitroprusside reaction, the so called Brand test, for detection of homocystine in urine (Chapter 1.5), has been performed in 16 out of the 21 studied homocystinuria patients. In our hands the results were rather erratic: in only 6 of the 16 homocystinurics this test was positive (> 20 mg/100 ml), in 4 there was a borderline result (10-20 mg/100 ml), and in as much as 6 patients the test was negative (< 10 mg/100 ml). The score of detection of the test might be increased by inducing excessive homocystinuria in the patients by means of a methionine loading[1]. However, it is noteworthy that in 6 out of the 10 patients with a negative or borderline result of the test, urine was screened excreted at the day of a standardized oral loading with 0.1 g/kg L-methionine. The occurrence of a false negative Brand test in 38% and only a borderline result in an additional 25% of these studied homocystinurics clearly demonstrates its complete unreliability. The modification of the test, described by Spaeth and Barber[7] as the silver-nitroprusside test which excludes the detection of cystinuria, may enhance the specificity, but not at all the sensitivity as these authors mistakenly claimed very recently[8].

After 6 weeks of treatment with pyridoxine, 750 mg daily, in the presence of plasma levels of folic acid in the normal range (Chapter 1.10.1), fasting levels of total homocysteine, methionine, and cystine nearly normalized in 14 patients, fell but remained still abnormal in 3 patients (patients nos 10, 11 and 19), and virtually did not change in patient no 16 (Table 2). In 3 patients the in

*Table 1*: Delay between the first signs and the clinical findings leading to the diagnosis of homozygous homocystinuria in 21 patients

| Patients | Sibships | Sex | First established signs | at age yr | age at diagnosis of homocystinuria yr | conditions leading to diagnosis of homocystinuria |
|---|---|---|---|---|---|---|
| no 1 | 1 | ♀ | pulmonary embolisms | 40 | 50 | ectopia lentis |
| | | | paranoid psychosis | 40 | | |
| no 2 | 2 | ♀ | ectopia lentis | 5 | 20 | pupillary block glaucoma |
| no 3 | 3 | ♀ | ectopia lentis | 4 | 20 | ectopia lentis |
| no 4 | 3 | ♂ | none | | 25 | sib with homocystinuria |
| no 5 | 3 | ♂ | seizures | 4 | 27 | sib with homocystinuria |
| | | | increasing myopia | 20 | | |
| | | | recurrent thrombophlebitis | 24 | | |
| | | | ectopia lentis | 26 | | |
| no 6 | 4 | ♀ | ectopia lentis | 19 | 21 | ectopia lentis |
| | | | pupillary block glaucoma | 20 | | |
| no 7 | 4 | ♀ | suicidal tendencies | 26 | 29 | sib with homocystinuria |
| | | | hypertension | 26 | | |
| | | | carotid arteries stenoses | 26 | | |
| no 8 | 5 | ♂ | increasing myopia | 7 | 22 | ectopia lentis |
| | | | marfanoid features | 8 | | marfanoid features |
| | | | mental retardation | 8 | | |
| | | | ectopia lentis | 16 | | |
| no 9 | 6 | ♀ | ectopia lentis | 17 | 20 | pupillary block glaucoma |
| no 10 | 7 | ♂ | ectopia lentis | 4 | 4 | ectopia lentis |
| no 11 | 7 | ♂ | scoliosis | ¼ | 3 | sib with homocystinuria |
| no 12 | 8 | ♀ | recurrent cerebrovascular accidents | 39 | 54 | cerebrovascular arteriosclerosis |
| | | | suicidal tendencies | 47 | | |
| no 13 | 9 | ♀ | deep venous crural thrombosis | 21 | 53 | ectopia lentis |
| | | | ectopia lentis | 36 | | retinal detachment |

| | | | | | | |
|---|---|---|---|---|---|---|
| no 14 | 9 | ♀ | angina pectoris | 24 | 50 | sib with homocystinuria |
| | | | puerperal psychosis | 24 | | |
| no 15 | 10 | ♀ | ectopia lentis | 7 | 31 | ectopia lentis |
| no 16 | 11 | ♂ | ectopia lentis | 1 | 15 | marfanoid features |
| | | | psychomotor retardation | 2 | | ectopia lentis |
| | | | scoliosis, marfanoid features | 6 | | |
| | | | oligophrenia | 6 | | |
| no 17 | 12 | ♂ | marfanoid features | 6 | 11 | ectopia lentis |
| no 18 | 13 | ♂ | ectopia lentis | 6 | 23 | ectopia lentis |
| no 19 | 14 | ♀ | psychomotor retardation | 7 | 9 | psychomotor retardation |
| | | | marfanoid features | 7 | | marfanoid features |
| no 20 | 15 | ♀ | ectopia lentis | 23 | 46 | psychotic depression |
| | | | fetal loss (6×) | 27-35 | | ectopia lentis |
| | | | depression | 34 | | |
| no 21 | 16 | ♂ | psychomotor retardation | 2 | 9 | deep venous crural thrombosis |
| | | | seizures | 3 | | peptic ulcer |
| | | | ectopia lentis | 6 | | oligophrenia |
| | | | pupillary block glaucoma | 8 | | ectopia lentis |
| | | 9♂ | | mean 13±3 SEM | 26±4 SEM | |
| | | 12♀ | | range ¼-40 | 3-54 | |

*Table 2*: Fasting serum levels of total homocysteine, methionine and cysteine (in µmol/l) before and during administration of pyridoxine, 750 mg daily for at least 6 weeks, in 21 homozygous homocystinuria patients

| Patients | Total homocysteine[£] | | Methionine | | Cystine | |
|---|---|---|---|---|---|---|
| | before pyridoxine | during pyridoxine | before pyridoxine | during pyridoxine | before pyridoxine | during pyridoxine |
| no 1 | 82 | 8 | 71 | 30 | 34 | 52 |
| no 2 | 62 | 9 | 80 | 33 | 30 | 58 |
| no 3 | 44 | 4 | 69 | 15 | 38 | 30 |
| no 4 | 122 | 18 | 52 | 51 | 20 | 58 |
| no 5 | 82 | 17 | 85 | 40 | 33 | 62 |
| no 6 | 189 | 6 | 81 | 34 | 10 | 43 |
| no 7 | 175 | 6 | 137 | 37 | 22 | 64 |
| no 8 | 14 | 3 | 34 | 24 | 48 | 51 |
| no 9[ff] | 274 | - | 48 | - | 6 | - |
| no 10[0] | 194 | 63 | 115 | 98 | 0 | 35 |
| no 11[0] | 73 | 58 | 156 | 70 | 0 | 38 |
| no 12[b] | 4 | 3 | 21 | 22 | 54 | 57 |
| no 13 | 156 | 15 | 83 | 30 | 19 | 56 |
| no 14 | 87 | 12 | 74 | 33 | 46 | 63 |
| no 15[b] | 7 | - | 35 | - | 52 | - |
| no 16[ff] | 204 | 174 | 549 | 655 | 4 | 6 |
| no 17 | 200 | - | 129 | - | 3 | - |
| no 18 | 42 | 10 | 62 | 51 | 38 | 51 |
| no 19 | 178 | 121 | 83 | 79 | 26 | 33 |
| no 20 | 263 | 14 | 71 | 30 | 23 | 68 |
| no 21 | +[&] | 25 | - | 59 | - | 46 |
| mean | 123±19SEM | 31±11 | 102±25 | 77±34 | 25±4 | 48±4 |
| range | 4 -274 | 3-174 | 21-549 | 15-655 | 0-54 | 6-68 |
| n = | 20 | 18 | 20 | 18 | 20 | 18 |

| | | | | |
|---|---|---|---|---|
| Control subjects | | | | |
| mean | 0.9±0.3SEM | premenopausal | 26±2 | 54±2 |
| range | 0.2-2.6 | women | 16-38 | 47-67 |
| n = 10 | | | | |
| mean | 2.6±0.4 | postmenopausal | 21±1 | 64±4 |
| range | 1.3-4.9 | women | 16-27 | 44-78 |
| n = 10 | | | | |
| mean | 3.3±0.3 | men | 28±2 | 61±2 |
| range | 1.1-5.6 | | 14-47 | 50-78 |
| n = 20 | | | | |

£ calculated as twice the concentration of homocystine plus the concentration of homocysteine-cysteine mixed disulphide (Chapter 1.3.1)
ff patients 1 and 2 in Chapter 4
° reference no 2
þ diagnosis based upon abnormal rise of total homocysteine after methionine loading and upon residual cystathionine synthase activity in cultured fibroblasts (Table 3)
& determined only qualitatively in urine in 1975

vivo response to pyridoxine administration is unknown, because in patient no 9 treatment was started only a few days before her death[6], patient no 15 refused this treatment, and in patient no 17 the effect of treatment could not be evaluated due to lack of facilities to do so in Kenya where he lives. However, the presence of residual cystathionine synthase activity in liver tissue from patient no 9 and in cultured fibroblasts in patient no 15 makes their in vivo pyridoxine responsiveness highly likely, as has been pointed out earlier (Chapter 1.4.3).

On the base of their biochemical in vivo response to pyridoxine treatment, Brenton and Cusworth[9] categorized homocystinuria patients in 3 groups: "completely" responsive, "partially" responsive, and non-responsive patients. According to this classification 14 patients (67%) out of our group of 21 homocystinurics were "complete" responders, 3 (14%) "partial" responders, and only 1 (5%) a non-responder to pyridoxine, whereas in 3 patients (14%) this response is unknown (Table 3). In the 629 homocystinuria patients surveyed by Mudd et al[3] these percentages were 37%, 10%, 37% and 16% respectively. The remarkably high frequency of pyridoxine responsiveness in our group of 21 homocystinurics may be due to the fact that in The Netherlands in contrast to in most other countries, screening for homocystinuria is not included in the newborn screening program. Indeed, the probability for a pyridoxine non-responsive patient to be detected by such screening is higher than for a responsive patient (Chapter 1.9). Moreover, undetected and thereby untreated pyridoxine non-responsive patients, because of their more severe symptoms (Chapter 1.6), are more likely to be institutionalized or to die at early age (Chapter 1.11) without being screened for the presence of homocystinuria.

The term "complete" pyridoxine responsiveness is in our opinion misleading. Pyridoxine treatment in "complete" responders nearly normalizes the aberrant amino acid profile only in the fasting state and only after minor protein intake. After a major methionine load however, marked homocysteinemia still occurs in these patients, as will be detailed in Chapter 3.

### 2.4 ENZYMATIC INVESTIGATIONS (Table 3)

Data on cystathionine synthase activity in the homocystinuria patients are given in Table 3. In all patients the enzyme activities have been measured in cultured fibroblasts from skin biopsies, except in patient no 9 in whom this activity was determined in liver tissue obtained at autopsy. In patient no 17 no enzyme assay was performed because of the lack of such facilities in Kenya.

In the cultured fibroblasts of the patients cystathionine synthase activity was either non-detectable (patients nos 7 and 16) or only residual, ranging from 0.01 to 0.37 nmol/h/mg protein. The mean residual activity was 0.07

nmol/h/mg protein, i.e. as low as 0.6% of mean activity in normal controls. In patient no 9 the enzyme activity in liver tissue was only 1.2% of the activity in a control liver obtained at autopsy. In patient no 15, a premenopausal woman, a somewhat higher residual activity of cystathionine synthase was measured, 0.37 nmol/h/mg protein, i.e. 3% of mean control activity. It is noteworthy that in the fasting state her level of total homocysteine was only minimally elevated (Table 2) and that this level showed the lowest rise after methionine loading from all patients (Table 3). The after load peak level of total homocysteine in fact was in the range of the levels found in obligate heterozygous premenopausal women (Chapter 7). Nevertheless, she is considered to be a homozygote because of her presenting symptom of ectopia lentis which never occurs in heterozygotes and her cystathionine synthase activity, which is low compared to the mean activity of 19% in obligate heterozygotes for homocystinuria (Chapter 7).
In postmenopausal patient no 12 suffering from severe cerebral arteriosclerosis and presenting a normal profile of amino acids in the fasting state (Table 2), the after load peak level of total homocysteine was in the homozygous range (Table 3) and clearly exceeded the range of levels obtained in obligate heterozygous postmenopausal women (Chapter 7). Cystathionine synthase activity in her fibroblasts undoubtedly was compatible with the homozygous condition. As additional proof of homozygosity, her 2 daughters appeared to be heterozygotes for homocystinuria (subjects nos 4 and 10 in Table 2, Chapter 7).
Patients nos 12 and 15 moreover illustrate that in exceptional homozygous homocystinuria patients fasting levels of total homocysteine, methionine, and cystine may be within the normal range and that such patients can only be unmasked as homozygotes by means of abnormally high levels of total homocysteine in response to methionine loading, in combination with low cystathionine synthase activity in their cultured fibroblasts.

In vivo responsiveness or non-responsiveness to pyridoxine administration, demonstrated in the patients in the foregoing paragraph, is related with the presence or absence of some residual activity of cystathionine synthase in their cultured fibroblasts (Table 3), according to the general rule as discussed in Chapter 1.4.3. As an exception, patient no 7 clearly in vivo responded to pyridoxine in the absence of any residual enzyme activity. Her sister, patient no 6 however, was also in vivo responsive but, she had still some residual enzyme activity. Thus, patient no 7 is one of the exceptional homocystinurics in whom it is demonstrated that the relation between in vivo responsiveness to pyridoxine and the presence of detectable residual enzyme activity is not an absolute one. On the other hand, this interesting observation in these homozygous sisters confirms the absolute concordance of responsiveness or non-responsiveness among patients within one sibship[3], as a reflection of the presence of a specific mutant cystathionine synthase within one family (Chapter 1.4.3).

*Table 3*: Cystathionine synthase activities, basal and after stimulation in vitro by addition of 1 mM pyridoxal phosphate, the in vivo responsiveness to pyridoxine, peak levels of total homocysteine after methionine loading and their decreases during pyridoxine administration for at least 6 weeks

| Patients | Sib-ships | Cystathionine synthase activity[£] | | Responsiveness in vivo to pyridoxine | Peak levels of total homocysteine after methionine loading[0] | |
|---|---|---|---|---|---|---|
| | | basal | after addition of 1mM pyridoxal phosphate[ff] | | before pyridoxine | during pyridoxine* |
| | | nmol/h/mg protein | | | µmol/l | |
| no 1 | 1 | 0.14 | +0.13 | 70 | 277 | -153 |
| no 2 | 2 | 0.04 | +0.06 | + | 177 | - 34 |
| no 3 | 3 | 0.03 | +0.15 | + | 131 | - 91 |
| no 4 | 3 | 0.03 | +0.07 | + | 194 | - 8 |
| no 5 | 3 | 0.03 | +0.08 | + | 192 | - 39 |
| no 6 | 4 | 0.05 | +0.05 | + | 202 | - 98 |
| no 7 | 4 | N.D.† | N.D. | + | 247 | -111 |
| no 8 | 5 | 0.09 | +0.28 | + | 159 | -114 |
| no 9 | 6 | 0.54[&] | +0.17 | unknown | - | - |
| no 10 | 7 | 0.02 | +0.03 | partial | 118 | - 28 |
| no 11 | 7 | 0.01 | +0.02 | partial | 112 | - 1 |
| no 12 | 8 | 0.09 | +0.05 | + | 104 | - 66 |
| no 13 | 9 | 0.06 | +0.03 | + | 290 | -146 |
| no 14 | 9 | 0.04 | +0.01 | + | 172 | - 80 |
| no 15 | 10 | 0.37 | +0.11 | unknown | 53 | - |
| no 16 | 11 | N.D. | N.D. | - | 215 | - 4 |
| no 17 | 12 | - | - | unknown | - | - |
| no 18 | 13 | 0.03 | +0.02 | + | 133 | - 39 |
| no 19 | 14 | 0.05 | +0.02 | partial | 184 | - 48 |
| no 20 | 15 | 0.01 | +0.01 | + | 342 | -220 |

| | | | | | | |
|---|---|---|---|---|---|---|
| no 21 | 16 | 0.05 | +0.23 | + | - | - |
| mean[b] | | 0.07±0.02SEM | +0.08±0.02SEM | 14+ | mean 183±17SEM | -75±15SEM |
| range | | 0.01-0.37 | +25%-+460% | 3 partial | range 53-342 | - 1--220 |
| | | | | 1- | | |
| | | | | 3 unknown | | |
| Control subjects | | | | | | |
| mean | | 11.3±1.1SEM | 14.2±1.5SEM | | mean 6.8±0.4SEM | premenopausal |
| range | | 3.6-25.7 | 5.0-32.1 | | range 5.0-9.1 | women |
| n = 23 | | | | | n = 10 | |
| | | | | | mean 24.1±1.9 | postmenopausal |
| | | | | | range 17.1-34.9 | women |
| | | | | | n = 10 | |
| | | | | | mean 13.3± 0.8 | men |
| | | | | | range 7.8–19.2 | |
| | | | | | n = 20 | |

£ in cultured fibroblasts; values are averages of at least 2 determinations
ff in vitro addition to the assay mixture
† not detectable
& in liver tissue obtained at autopsy; level in control liver 45.7 before and 66.8 after addition of 1 mM pyridoxal phosphate to the assay mixture
° maximal achieved levels between 4 to 8 hours after oral methionine loading (0.1 g/kg)
þ mean residual activity and mean of the increases after addition of 1 mM pyridoxal phosphate in cultured fibroblasts
* differences between peak levels of total homocysteine after methione loading before and during treatment with pyridoxine

In this group of homozygous homocystinuria patients there was a weak though statistically significant correlation between their basal cystathionine synthase activities and their fasting levels of total homocysteine in serum shown in Table 2 ($r_s = -0.56$; $P < 0.05$). Although intelligible, such correlation has never been reported before.

Peak levels of total homocysteine after a major methionine loading may be considered to reflect the capacity of the transsulphuration pathway if maximally stressed[12]. The procedure of a standardized oral methionine loading test as performed in 18 out of the 21 homozygous homocystinuria patients, is described in the "Methods" section of Chapter 3. The peak levels of total homocysteine achieved between 4 to 8 hours after loading with 0.1 g/kg L-methionine in these patients are given in the right panel of Table 3. They ranged from 53 to 342 µmol/l. There was no statistically significant correlation between the after load peak levels of total homocysteine and the basal cystathionine synthase activities in these patients ($r_s = -0.03$; $P > 0.10$).

In Chapter 1.4.3 it is pointed out that residual cystathionine synthase activity in cultured fibroblasts is stimulated by in vitro addition of 1 mM pyridoxal phosphate to the assay mixture. Enhancement varying from 30% to 1000% has been reported. Table 3 shows the results of such in vitro stimulation in the studied homozygous patients. Increases of enzyme activity ranging from 25% to as high as 460% were measured. The in vivo reponse of the patients to pyridoxine administration, 750 mg daily for at least 6 weeks, is shown in Table 3 in terms of the decrease of peak levels of total homocysteine after a methionine load similar as before the treatment with pyridoxine. No statistically significant correlation between the in vitro enhancement of the enzyme activities by pyridoxal phosphate and the in vivo decrease of after load peak levels of total homocysteine by treatment with high-dose pyridoxine was established ($r_s = 0.18$; $P > 0.10$).

### 2.5 TREATMENT (Table 4)

A systematic and prospective study of the biochemical and clinical effects of a number of treatment modalities we have not started as yet. Nevertheless, a number of pilot observations allows a preliminary account. Six weeks after the start of each therapy regimen the biochemical response was evaluated by amino acid analysis in fasting serum. In 19 out of the 21 homocystinuria patients treatment was started immediately after making the biochemical diagnosis of homocystinuria due to cystathionine synthase deficiency (Table 4). Patient no 9 died after only 2 days of treatment[6]. Patient no 14 refused any therapy until two years after diagnosis when recurrent retinal detachments occurred. Patient no 15 rejects treatment up to the present.

Table 4 shows that *pyridoxine* (Chapter 1.10.1) was prescribed to all patients, in a dose of 750 mg daily in most of them. This high dose proved maximally effective both on fasting (Table 2) and on peak levels of total homocysteine after methionine loading (Table 3). In patients nos 10 and 11 the effect of a daily dose of 500 mg pyridoxine was similar to that of 750 mg. In patients nos 19 and 21 the effect of pyridoxine as the sole therapy has been evaluated only with doses of 500 mg and 250 mg daily respectively. Thereafter, the dose was increased to 750 mg daily and folic acid was given concomitantly. In patient no 8 it was possible to reduce the daily dose of pyridoxine to as low as 90 mg without jeopardizing the benificial effect. The marked response to pyridoxine in this patient is reminiscent of that in some reported homocystinurics as reviewed by Mudd and Levy[13]. They showed responses to doses of pyridoxine as low as 10 and even 2 mg daily.

In 12 of the 19 treated homocystinuria patients treatment with pyridoxine remained the sole therapy, because of near normalization of the fasting amino acid levels in these patients. In patient no 17 pyridoxine was prescribed without the possibility of close follow-up evaluation.

In 5 patients *folic acid* therapy was given in addition to pyridoxine (Chapter 1.10.3). Patient no 21, considered as a "complete" responder, suffered from increasingly frequent seizures despite treatment with anticonvulsive drugs and pyridoxine. In an attempt to control these events, a further normalization of the amino acid profile was aimed by means of folic acid administration. Suchlike treatment had no biochemical effects as shown in Table 4, although seizures became less frequent.

The three "partially" pyridoxine responsive patients also were treated with additional folic acid. As shown in Table 4, this treatment did not lead to a substantial further decrease of fasting homocysteinemia in the patients nos 10 and 11. Methionine levels rose in them after starting with folic acid probably as a result of enhancement of re-methylation of homocysteine to methionine by folic acid excess. In patient no 19 additional folic acid treatment clearly lowered homocysteinemia. This effect may partially be accounted by the simultaneously prescribed higher dose of 750 mg pyridoxine. Nevertheless, a rather high level of homocysteine persisted in this patient.

Patient no 16 was considered pyridoxine non-responsive. After addition of folic acid the homocysteine level decreased but still remained as high as 117 µmol/l (Table 4). Moreover, the methionine levels markedly rose up to 974 µmol/l.

A *methionine-restricted diet* in addition to treatment with pyridoxine and folic acid did not have a homocysteine-lowering effect in patient no 10[2]. Therefore, it was discontinued after about half a year. In patient no 16, however, the dietary restriction of 600 mg methionine daily led to a further decrease of the homocysteinemia whereas the methionine level dropped from 974 to 118 µmol/l. Cystine concentration rose probably due to cystine supplements in the used amino acid mixture. This diet was well accepted by the patient and is continued up to the present in addition to folic acid

*Table 4*: Treatment in 21 homozygous homocystinuria patients

| Patients | Pyridoxine | | | + Folic acid | | | + Methionine restriction | | | + Betaine | | | Aceno-coumarol |
|---|---|---|---|---|---|---|---|---|---|---|---|---|---|
| | start | dose mg daily* | effect | start | dose mg daily | effect | start | mg daily | effect | start | dose g daily | effect | start |
| no 1 | 1979 | 750 | see table 2 | – | | | – | | | – | | | 1979 |
| no 2 | 1978 | ” | ” | – | | | – | | | – | | | 1978 |
| no 3 | 1978 | ” | ” | – | | | – | | | – | | | 1978 |
| no 4 | 1979 | ” | ” | – | | | – | | | – | | | 1979 |
| no 5 | 1978 | ” | ” | – | | | – | | | – | | | 1978 |
| no 6 | 1977 | ” | ” | – | | | – | | | – | | | 1977 |
| no 7 | 1978 | ” | ” | – | | | – | | | – | | | 1975 |
| no 8 | 1968 | 90 | H⁰ 14 → 4<br>M 34 → 32<br>C 48 → 67 | – | | | – | | | – | | | 1968-1975<br>>1980 |
| no 9 | – | | | – | | | – | | | – | | | – |
| no 10 | 1970 | 500 | H 194 → 63<br>M 115 → 98<br>C 0 → 35 | 1971 | 5 | H 63 → 53<br>M 98 → 135<br>C 35 → 19 | 1971 | 360 | "no effect" stop after ½ year | 1985 | 6 | H 105 → 27<br>M 186 → 131<br>C 36 → 45 | – |
| no 11 | 1971 | 500 | H 73 → 58<br>M 156 → 70<br>C 0 → 38 | 1971 | 5 | H 58→ 52<br>M 70 → 169<br>C 38 → 22 | – | | | 1985 | 6 | H 58 → 29<br>M 96 → 128<br>C 52 → 52 | – |
| no 12 | 1981 | 750 | see table 2 | – | | | – | | | – | | | 1981 |
| no 13 | 1982 | 750 | H 156 → 15<br>M 83 → 30<br>C 19 → 56 | – | | | – | | | – | | | 1982[xx] |
| no 14 | 1984 | 750 | see table 2 | – | | | – | | | – | | | – |
| no 15 | – | | | – | | | – | | | – | | | – |

| | | | | | | | | | | | | | |
|---|---|---|---|---|---|---|---|---|---|---|---|---|---|
| no 16 | 1982 | 750 | H 204 → 174<br>M 549 → 655<br>C 4 → 6 | 1982 | 15 | H 174 → 117<br>M 655 → 974<br>C 6 → 8 | 1983 | 600 | H 117 → 84<br>M 974 → 118<br>C 8 → 31 | 1985 | 6 | H 84 → 13<br>M 118 → 477<br>C 31 → 50 | 1982[£] |
| no 17 | 1983 | 750 | ? | – | | | – | | | – | | | – |
| no 18 | 1984 | 750 | see table 2 | – | | | – | | | – | | | – |
| no 19 | 1984 | 500 | H 178 → 121<br>M 83 → 79<br>C 26 → 33 | 1984 | 5[þ] | H 121 → 62<br>M 79 → 80<br>C 33 → 41 | – | | | 1985 | 6 | H 62 → 18<br>M 80 → 69<br>C 41 → 46 | – |
| no 20 | 1984 | 750 | see table 2 | – | | | – | | | – | | | – |
| no 21 | 1975 | 250 | H "+" → 25<br>M ? → 59<br>C ? → 46 | 1984 | 5[þ] | H 25 → 29<br>M 59 → 58<br>C 46 → 44 | – | | | – | | | – |

* minimal dose with maximal effect

° H = total homocysteine, M = methionine, C = cystine, in fasting serum

þ in combination with 750 mg pyridoxine daily

xx stopped in 1983 because of bleeding duodenal ulcer

£ stopped in 1985; since 1985 dipyridamole, 225 mg daily

treatment. In spite of the minor decrease of the homocysteinemia after pyridoxine treatment in this patient pyridoxine was continued yet on the strength of reports in literature stressing the beneficial effects of pyridoxine even in non-responsive patients[14-16].

In those patients in whom pyridoxine in combination with folic acid (patients nos 10, 11 and 19) and with a methionine restricted diet (patient no 16) did not induce near normalization of the amino acid levels, we started therapy with ***betaine*** in addition since January 1985. In all these patients the methyl-donor betaine (Chapter 1.10.3) in a daily dose of 6 grams, led to a substantial decrease of the homocysteinemia (Table 4). A major increase of methionine levels, as a sign of the enhanced rate of re-methylation of homocysteine to methionine, was observed only in patient no 16. This induction of hypermethioninemia will require careful evaluation of this treatment in the future[3], although early reports on methionine toxicity in animals have not been confirmed in men[13].

The observation of severe vascular abnormalities in young patients out of this group of homozygous homocystinurics has led to the treatment with the anticoagulant acenocoumarol as a preventive measure even in the pyridoxine responders (Chapter 1.10.3)[1]. In young patients treated with pyridoxine with or without folic acid from early ages on and not presenting clinical signs of major vascular complications (patients nos 10, 11, 18, 19, 21), such therapy was not started. Patients nos 14 and 20 refused acenocoumarol treatment. In patient no 16 the drug was prescribed when major surgery because of severe scoliosis was performed and longstanding immobilization in a plaster brace was indicated (Chapter 4, patient no 2). No thromboembolic complications did occur in this severely affected patient. When additional betaine therapy had proven to induce a substantial further decrease of homocysteinemia (Table 4), the strict preventive measure of acenocoumarol treatment was stopped and since then the platelet aggregation inhibiting agent dipyridamole (Chapter 1.10.3), 225 mg daily, is prescribed.

## 2.6 CLINICAL FINDINGS AT DIAGNOSIS AND SINCE THE START OF THERAPY

### *2.6.1 Ocular abnormalities (Table 5)*[17]

Ectopia lentis, always bilateral, has been established before or at the time of diagnosis of homocystinuria in 16 (76%) out of the 21 homozygous patients. In 4 of them it had already resulted in bilateral and in one in unilateral aphakia because of surgical lens extraction prior to the diagnosis of homocystinuria. In all patients ectopia lentis had been accompanied by lenticular myopia which is due to the increased curvature of the lens after disruption of the zonulae fibres (Chapter 1.6.1). By ongoing loosening of the lens such myopia may be progressive which is clearly demonstrated in the morbid history of several of these patients (Table 5). From the moment that – due to maximal subluxation of the lens – the aphakic part of the pupil is used for vision, the refraction is no longer myopic. This condition of "aphakic vision in ectopia lentis" was present in 4 patients at diagnosis. In only one homocystinuric, patient no 7, high myopia was established in the absence of ectopia lentis.

Pupillary block glaucoma, an acute painful complication of ectopia lentis caused by luxation of the lens into the anterior chamber, leading to surgical lens extraction, occurred in 6 patients before establishing homocystinuria. Such event has been the reason for referral to the Ophthalmological Department of this hospital and finally led to diagnostic manoeuvres to detect homocystinuria. In patient no 13 the complication of retinal detachment, 17 years after lens extraction because of ectopia lentis, led to the diagnosis.

It is of note that after the start of therapy for homocystinuria ectopia lentis did not occur during follow-up periods ranging from ½ to 13 years in the 5 patients who did not present this symptom at the time of diagnosis. These patients all were pyridoxine responders. According to the probability score to have no ectopia lentis at certain ages in untreated though pyridoxine responsive homocystinurics calculated by Mudd et al[3], such probabilities for 3 of our 5 patients were less than 10%, in patient no 11 17%, and in patient no 19 45%. These data seem to confirm the conclusion of Mudd that pyridoxine treatment in responsive patients prevents the occurrence of ectopia lentis[3]. However, longer follow-up periods for definitive evaluation of the effect of treatment are needed in these patients.

In the 16 patients suffering from ectopia lentis before or at the time of the diagnosis of homocystinuria major complications did not occur after the start of therapy for periods ranging from ½ to 16 years, except for retinal detachment in patient no 3 six years after the start with pyridoxine and in patient no 14 two years after diagnosing homocystinuria. The latter patient had refused therapy until this complication occurred. In patient no 8 pupillary block glaucoma happened very shortly after the start with pyridoxine.

The frequency of ectopia lentis in this group of homocystinuria patients is about equal to the 82% established by Mudd et al[3] in 629 homocystinurics. It is remarkable that in patient no 16, the only pyridoxine non-responsive patient among our homocystinurics, ectopia lentis had been established as early as at the age of one year, whereas the youngest patient reported in literature with ectopia lentis was 1½ years old[18]. Nevertheless, there has been a delay in making the correct diagnosis in this patient until the age of 15 due to the wrong diagnosis of Marfan's syndrome, as will be discussed in detail in Chapter 4.

*Table 5*: Ocular abnormalities in 21 homozygous homocystinuria patients before and at the diagnosis of homocystinuria and after the start of therapy

| Patients | Before diagnosis | | at diagnosis | | after start of therapy | |
|---|---|---|---|---|---|---|
| no | year(age) | abnormalities* | year(age) | abnormalities | start | changes |
| 1 | | High myopia** | 1979(50) | High myopia | 1979 | Lens extraction LE 1979 |
| | 1978(49) | Ectopia lentis | | Ectopia lentis | | Lens extraction RE 1981 |
| 2 | | High myopia | 1978(20) | Aphakic vision*** | 1978 | Lens extraction LE 1978 |
| | 1963( 5) | Ectopia lentis | | in ectopia lentis | | |
| | | | | Pupillary block glaucoma LE | | |
| 3 | 1962( 4) | High myopia | 1978(20) | Aphakia | 1978 | Retinal detachment RE 1985 |
| | 1963( 5) | Ectopia lentis | | | | |
| | 1965( 7) | Lens extraction RE + LE | | | | |
| 4 | No ocular | abnormalities | 1978(25) | No | 1979 | No |
| 5 | 1971(20) | Increasing myopia | 1978(27) | High myopia | 1978 | No |
| | 1977(26) | Ectopia lentis | | Ectopia lentis | | |
| 6 | | High myopia | 1977(21) | Ectopia lentis LE | 1977 | Lens extraction LE 1977 |
| | 1975(19) | Ectopia lentis | | Pupillary block glaucoma LE | | |
| | 1976(20) | Pupillary block glaucoma RE | | Aphakia RE | | |
| | | & lens extraction RE | | | | |
| 7 | 1975(26) | High myopia | 1977(28) | High myopia | 1978 | Increased myopia 1983 |
| 8 | 1953( 7) | Increasing myopia | 1968(22) | Aphakic vision | 1968 | Pupillary block glaucoma LE 1969 |
| | 1962(16) | Ectopia lentis | | in ectopia lentis | | Lens extraction LE 1969 |
| 9 | 1974(14) | Increasing myopia | 1980(20) | High myopia | Deceased | in 1980 |
| | 1977(17) | Ectopia lentis | | Ectopia lentis | | |
| | | | | Pupillary block glaucoma RE | | |
| | | | | Vitreous haemorrhage RE | | |
| 10 | 1970( 4) | High myopia | 1970( 4) | High myopia | 1970 | Aphakic vision in ectopia |
| | | | | Ectopia lentis | | lentis 1979 |
| 11 | No ocular | abnormalities | 1970( 3) | No | 1971 | No |
| 12 | No ocular | abnormalities | 1981(54) | No | 1981 | No |

| | | | | | | |
|---|---|---|---|---|---|---|
| 13 | | Increasing myopia | 1982(53) | Aphakia | 1982 | No |
| | 1964(36) | Ectopia lentis | | Retinal detachment RE | | |
| | 1965(37) | Lens extraction RE + LE | | | | |
| 14 | 1933( 1) | Enucleation LE | 1982(50) | High myopia RE | 1984 | Retinal detachment RE 1984 |
| | | Increasing myopia RE | | Ectopia lentis RE | | |
| | | | | Glaucoma RE | | |
| | | | | Cataract RE | | |
| 15 | 1957( 7) | Increasing myopia | 1981(31) | Aphakic vision in ectopia lentis | No | treatment |
| | 1957( 7) | Ectopia lentis | | | | |
| | 1959( 9) | Aphakic vision in ectopia lentis | | | | |
| 16 | 1968( 1) | Ectopia lentis | 1982(15) | Aphakic vision in ectopia lentis | 1982 | No |
| | | Myopia? | | | | |
| | | | | Cataract | | |
| 17 | | Myopia? | 1982(11) | Ectopia lentis | | |
| | | | | Pupillary block glaucoma RE | 1982 | ? |
| 18 | 1967( 6) | High myopia | 1984(23) | High myopia | 1984 | No |
| | 1967( 6) | Ectopia lentis | Ectopia lentis | | | |
| 19 | No ocular | abnormalities | 1984( 9) | No | 1984 | No |
| 20 | 1952(14) | Increasing myopia | 1984(46) | Aphakia | 1984 | No |
| | 1961(23) | Ectopia lentis | | Corneal dystrophy | | |
| | 1962(24) | Pupillary block glaucoma LE & lens extraction LE + RE | | | | |
| 21 | | Myopia? | 1975( 9) | Aphakia | 1975 | No |
| | 1972( 6) | Ectopia lentis | | | | |
| | 1972( 6) | Pupillary block glaucoma LE & lens extraction LE | | | | |
| | 1974( 8) | Pupillary block glaucoma RE & lens extraction RE | | | | |

* all abnormalities are bilateral, except if indicated RE (right eye) or LE (left eye)

** high myopia: 5 diopters or more

*** aphakic vision: for vision the aphakic part of the pupil is used because of maximal subluxation of the lens

*2.6.2 Skeletal abnormalities (Table 6)*

Table 6 shows that osteoporosis was established in 11 (65%) of the 17 homocystinuria patients in whom roentgenographic examination of the spine has been performed at the time of the diagnosis of homocystinuria. In 10 out of these 17 patients also histopathological studies of bone tissue obtained at biopsy of the iliac crest have been performed. In 7 patients the result of this latter examination confirmed the roentgenographic presence or absence (patient no 5) of osteoporosis. In 3 patients however, the histopathological study revealed signs of osteoporosis whereas roentgenographic examination of the spine failed to do so. In 11 patients results of such histopathological examination at diagnosis of homocystinuria are not available. Therefore, combining the roentgenographic and histopathological data we established a frequency of osteoporosis at least as high as 82% in these homocystinurics. This figure is in accordance with data from literature (Chapter 1.6.2). However, in all of the patients, except one (patient no 6) the osteoporosis was mild. In none of the patients osteoporosis had led to pathological fractures or vertebral collapse.

Scoliosis was present at diagnosis in 17 (81%) of the 21 patients and in the majority of them roentgenographic examinations were available confirming the clinical abnormality. In most cases the scoliosis was moderate before and at the time of the diagnosis of homocystinuria, except in patient no 11 in whom treatment with a plaster brace was necessary already at the age of 3 months, and in patient no 16 in whom surgical correction had to be performed at age of 16 years soon after the diagnosis of homocystinuria (Chapter 4, patient no 2). The frequency of scoliosis (81%) is higher than the about 50% reported in literature (Chapter 1.6.2). Most probably, in those studies mild cases of scoliosis have not been taken into account. In our group of patients scoliosis was without exception accompanied by the roentgenographic sign of biconcavely flattened vertebrae, the so-called fish-vertebrae (Chapter 1.6.2, Figure 2).

In 13 (62%) patients marfanoid features were found, either dolichostenomelia (9 patients), i.e. excessively lengthened long bones (Chapter 1.6.2), or arachnodactyly (11 patients), or both (7 patients). Arachnodactyly was considered to be present in the case of a length-to-width ratio of the 2nd to 5th metacarpal bones exceeding the mean + 3 S.D. in normal controls[19] (Figure 1). In the 13 patients the marfanoid features were not accompanied by excessive height in all as only 4 of them exceeded the 90th percentile (patients nos 2, 9, 18 and 19). The height of patient no 4 was above this percentile in the absence of marfanoid features.

Marked genu valgum was found in the patients nos 10, 11 and 16, pectus carinatum in the patients nos 9 and 16, and pectus excavatum in patients nos 1 and 19. High, so called gothic palate was a sign in patients nos 9, 16, 18 and

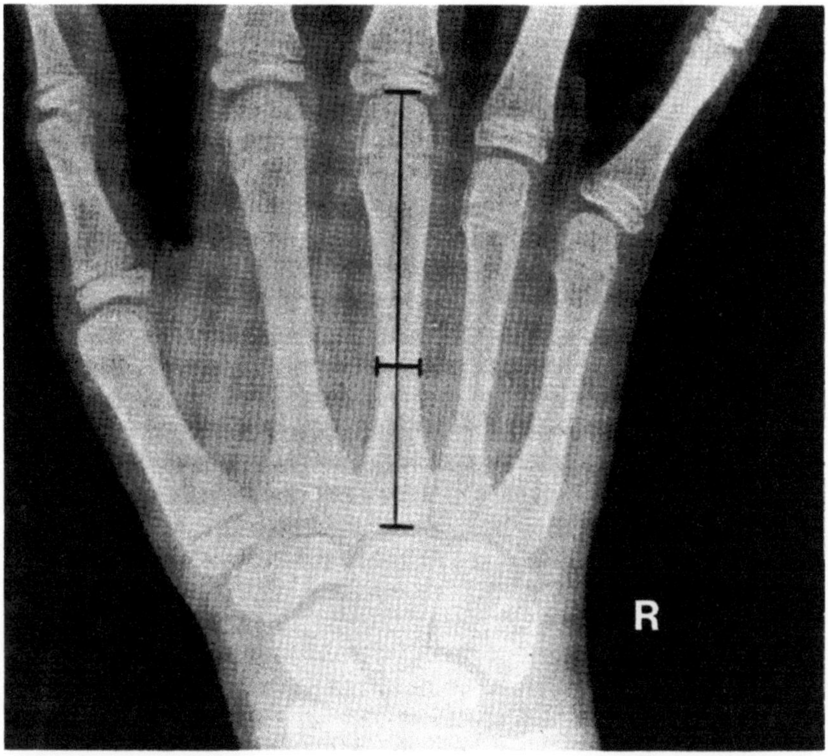


*Fig. 1.* Arachnodactyly in patient no 19: metacarpal index (length-to-width ratio of the 2nd to 5th metacarpal bones) is 10.13 versus 7.60 ± 0.52 S.D. in the right hand of a normal female.

19. Patient no 19 also showed marked protrusion of the upper teeth. Pes cavus was observed in the patients nos 9, 16 and 19, pes planus in patients nos 10, 11 and 20. Patients nos 16, 19 and 21 walked with the typical "Charlie-Chaplin"-gait.

Among the 12 patients in whom osteoporosis could be evaluated after the start of treatment for periods ranging from 2 to 14 years, this abnormality at yearly roentgenographic examinations with or without simultaneously performed histopathological studies remained at least unchanged in 5 of them, decreased in another 5, and stayed absent in one, patient no 5. In homocystinurics like the latter patient, a pyridoxine responder at age of 33 years, Mudd et al[3] calculated a probability score of having radiographic spinal osteoporosis of at least 74%. Although the data in our group of patients are suggestive of some beneficial effect of treatment of homocystinuria on preventing or ameliorating osteoporosis in homocystinurics, a solid conclusion awaits

*Table 6*: Skeletal abnormalities at diagnosis and after start of therapy in 21 homozygous homocystinuria patients

| Patients | At diagnosis of homocystinuria | | | | | | | After start of therapy | | | | |
|---|---|---|---|---|---|---|---|---|---|---|---|---|
| | year(age) | osteoporosis* | | scoliosis | height cm | marfanoid features | | start | osteoporosis | | scoliosis | height cm |
| | | roent-geno-graphic | histo-patho-logical | | | dolicho-stenome-lia** | arachno-dactyly⁰ | | roent-geno-graphic | histo-patho-logical | | |
| no 1 | 1979(50) | - | + | + | 168 | - | - | 1979 | - | ↓ | = | |
| no 2 | 1978(20) | + | + | + | 175 | + | + | 1978 | = | ↓ | = | |
| no 3 | 1978(20) | - | + | + | 163 | - | + | 1978 | - | ? | = | |
| no 4 | 1978(25) | - | + | + | 190 | - | - | 1979 | - | ? | = | |
| no 5 | 1978(27) | - | - | + | 183 | - | + | 1978 | - | - | = | |
| no 6 | 1977(21) | + | + | + | 170 | + | - | 1977 | = | ↓ | = | |
| no 7 | 1977(29) | + | + | - | 163 | - | - | 1978 | = | ↓ | - | |
| no 8 | 1968(22) | + | ? | + | 176 | + | + | 1968 | ->1980 | +1980 | = | |
| no 9 | 1980(20) | ? | ? | + | 185 | + | - | deceased in 1980 | | | | |
| no 10 | 1970( 4) | + | ? | +1975 | 105 | - | - | 1970 | = | ? | = | 184(18) |
| no 11 | 1970( 3) | + | ? | +1968 | 102 | - | - | 1971 | ↓ | ? | ↓↓ | 185(17) |
| no 12 | 1981(54) | ? | ? | - | 168 | - | - | 1981 | ? | ? | - | |
| no 13 | 1982(53) | + | + | + | 162 | - | + | 1982 | = | = | = | |
| no 14 | 1982(50) | + | ? | - | 167 | - | + | 1984 | | | | |
| no 15 | 1981(31) | - | ? | - | 173 | - | - | no treatment | | | | |
| no 16[b] | 1982(15) | + | ? | +1973 surgical correction in 1983 | 187 | + | + | 1982 | = | ? | =>1983 | |

| | | | | | | | | | | | | |
|---|---|---|---|---|---|---|---|---|---|---|---|---|
| no 17 | 1982(11) | ? | ? | + | ? | + | + | 1982 | ? | ? | ? | |
| no 18 | 1984(23) | + | + | + | 190 | + | + | 1984 | | | | |
| no 19 | 1984( 9) | + | ? | + | 158 | + | + | 1984 | | | | |
| no 20 | 1984(46) | – | ? | + | 161 | – | – | 1984 | | | | |
| no 21 | 1975( 9) | ? | ? | + | 152 | + | + | 1975 | ? | ? | ↑ ↑ | 172(19) |

* roentgenographic osteoporosis of the spine; histopathological osteoporosis in bone obtained at biopsy of the iliac crest

** excessive length of long bones with ratio upper segment/lower segment less than 0.9

° roentgenographic metacarpal index (lenght-to-width ratio of 2nd to 5th metacarpal bones of right hand) exceeding the mean + 3 S.D. in normal men or women (Figure 1)[19]

þ see Chapter 4, patient no 2, Figures nos 4 and 5.

more longitudinal studies in a greater number of early treated patients (Chapter 1.11).

In 11 out of the 17 patients with scoliosis established before or at the time of the diagnosis of homocystinuria, yearly roentgenographic evaluations of the spine after the start of treatment are available up to the present for periods ranging from 2 to 14 years. In 10 of them scoliosis remained unchanged and in patient no 11 it decreased. In patient no 21 it showed a severe increase at physical examination. Two of the 4 patients not presenting scoliosis at diagnosis of homocystinuria, could be yearly evaluated since the start of their treatment. In these patients no scoliosis developed for periods of 3 and 6 years. These data on the follow-up evaluation of scoliosis should be considered in relation with the above mentioned data on osteoporosis after the start of therapy, as in homocystinuria the former abnormality is generally held to be caused by the latter. Indeed, in the patients showing an unchanged or decreased osteoporosis after the start of treatment, scoliosis also remained unchanged or had decreased.

### *2.6.3 Vascular pathology (Table 7)*

In 18 out of the 21 homozygous homocystinuria patients major or minor involvement of the vascular system had been established before or at the time when the diagnosis of homocystinuria was made.

In 7 patients severe vascular pathology was present already at young age whereas only in one of them, patient no 12, the occurrence of multiple vascular events had been the cause of investigation for the presence of homocystinuria. The vascular complications in these 7 homocystinurics (patients nos 6, 7, 8, 9, 12, 14 and 21) will be discussed to some detail below.

*Patient no 6*, a woman who was referred at age of 21 because of ectopia lentis, presented at diagnosis of homocystinuria hypertension (160/100 mmHg), ischemia of the 3rd and 4th toe of the left foot, and an abdominal bruit. Selective angiographic examinations revealed areas of severe narrowing and small aneurysms of interlobar arteries of both kidneys (Figure 2), in the common hepatic, the left gastric and the gastroduodenal arteries. The main stem of the superior mesenteric artery also was narrowed whereas some of its branches were stenosed or occluded more distally (Fig. 3). No splenic artery could be identified; however, many arterial collateral vessels supplied blood to the spleen[20]. In addition, Doppler flow examination of cerebral and peripheral arteries revealed an abnormal flow velocity in the left femoral artery. Despite these striking vascular abnormalities, she had no ischemic complaints and no vascular events happened. Since the start of treatment with pyridoxine and acenocoumarol in 1977 such events did not occur as yet. Blood pressure is still moderately elevated (150/100 mmHg), the abdominal bruit and the acrocyanosis of the toes have resolved, and yearly performed

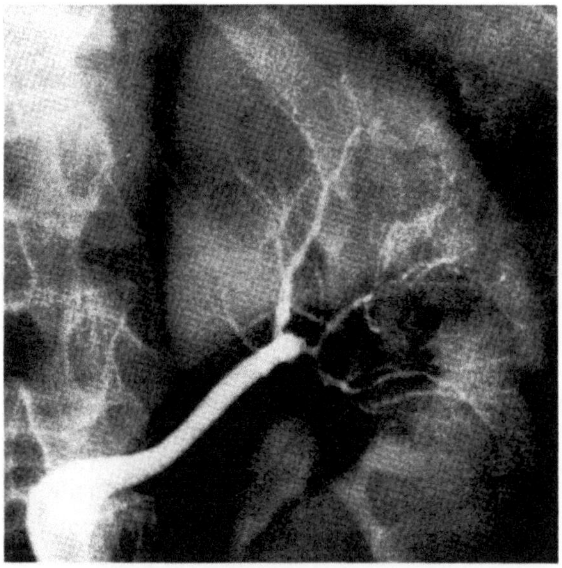

*Fig. 2.* Selective arteriogram of the upper of the two left renal arteries in patient no 6; note irregularity of many proximal and peripheral arteries (the lower left renal artery showed similar changes).

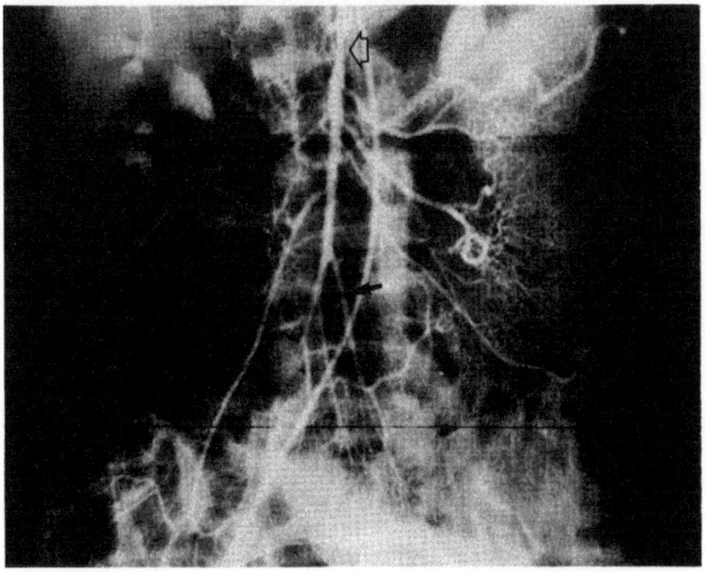

*Fig. 3.* Superior mesenteric arteriography in patient no 6; note the narrowing of the main stem of the superior mesenteric artery (open arrow), stenosis of a peripheral branch (large black arrow), and complete occlusion of one branch (between small black arrows).

Doppler flow examinations showed normal characteristics also of the left femoral artery since 1980 (Table 7).
Her sister, *patient no 7*, suffered from an organic psychosyndrome characterized by paranoid depression and attempted suicide at the age of 26 after birth of her second child, 3 years before homocystinuria was established. Physical examination at that time showed hypertension (140/100 mmHg) and transient neurologic abnormalities such as nystagmus, hemiparesis, and hemihypaesthesia. Angiographic examination revealed fibromuscular dysplasia of the carotid arteries with severe narrowing of the right internal carotid artery. Acenocoumarol as a preventive anticoagulant remedial was started. In 1977 investigations for the presence of homocystinuria were performed after establishing this diagnosis in her sister (patient no 6). Homocystinuria indeed was diagnosed in her. Angiographic re-examination showed irregularities and narrowing of both internal carotid arteries with severe stenosis of the right internal carotid artery. Abdominal angiography demonstrated an irregular caliber of the splenic artery with an aneurysm (Figure 4)[20]. Doppler flow examination also showed abnormal characteristics of the right internal carotid artery. After the start of additional treatment with pyridoxine in 1978, blood pressure remained moderately elevated (140/100 mmHg), no vascular events happened until the present and the psychosyndrome resolved completely. Doppler flow characteristics of the right carotid artery were less abnormal in 1983, whereas in 1984 a normal Doppler flow was obtained.
In *patient no 8* a cerebrovascular accident occurred at the age of 22 leading to hemiplegia, focal epilepsy, and apnea, soon after the admission to another hospital because of ectopia lentis. Homocystinuria was diagnosed at that time. After one month of bed rest calf vein thrombosis developed. Since then therapy with pyridoxine in combination with acenocoumarol was prescribed. Because of the uneventful course anticoagulant treatment was stopped in 1975. When he was referred to our hospital in 1980, moderate hypertension was established (140/100 mmHg). Doppler flow examination showed abnormal characteristics of one vertebral artery and acenocoumarol was prescribed again since then. At yearly examinations Doppler flow abnormalities were no longer found since 1983. No further vascular events happened up to the present.
The morbid history of *patient no 9* will be presented in detail in Chapter 4. She suffered from ectopia lentis since the age of 17 when the wrong diagnosis of Marfan's syndrome was made. From the age of 18 she had complaints of severe migraine. A few months after the start with oral contraceptives – which are contraindicated in the case of homocystinuria (Chapter 1.10.3) –, she developed acute pupillary block glaucoma necessitating admission to our hospital where investigations for the presence of homocystinuria were started immediately. However, 5 days after referral she had a hemiplegia and developed a comatous state. Papiledema was established. Angiographic examina-

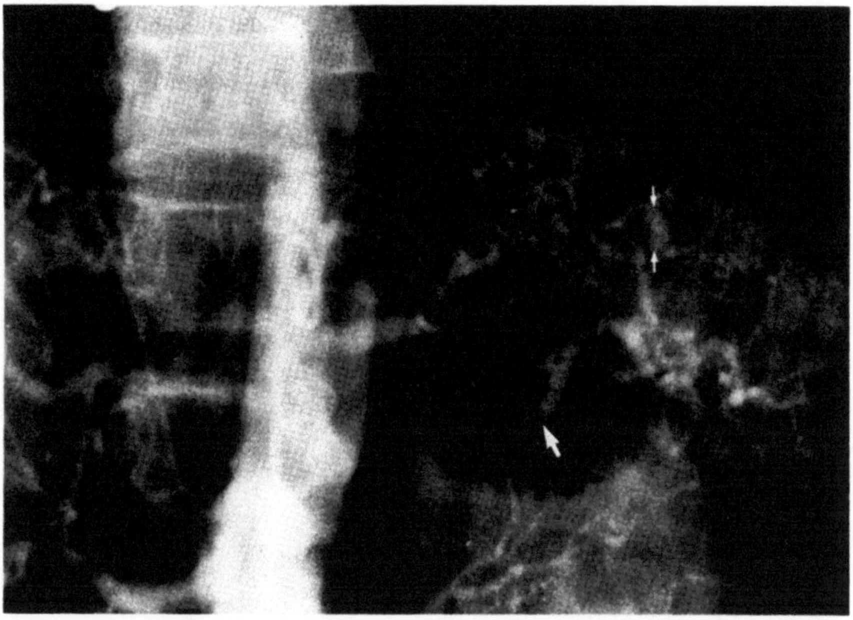

*Fig. 4.* Abdominal aortography in patient no 7; the splenic artery is irregularly shaped (large arrow), and a splenic aneurysm is present (between small arrows).

tion showed extreme narrowing of the common and the cervical internal and intracranial right carotid artery. There was an elevation of the lateral part of the middle cerebral artery and of the Sylvian triangle, whereas the anterior cerebral artery and the internal cerebral vein were displaced towards the left (Figures 5 and 6). No occlusions were visible, but the blood flow was so slow that no filling of the cortical veins or the superior sagittal sinus was obtained[6]. A CT-scan showed signs of temporal infarction, and the results of both investigations together raised the possibility of temporal infarction with brain edema. Meanwhile the biochemical diagnosis of homocystinuria due to cystathionine synthase deficiency was made and therapy with pyridoxine was started parenterally. Unfortunately, epileptic state and respiratory failure developed and the patient died 3 days after the onset of coma. Postmortem examination revealed signs of generalized thrombosis[6]. Recent infarctions of the renal parenchyma, the cardiac septum, and the myocardium of the left ventricle were seen. There was a haemorrhagic infarction in the right superior temporal gyrus with signs of edema in the surrounding areas. A small intact aneurysm was seen in the right middle cerebral artery. Multiple thrombi were observed in the superior sagittal sinus and in several leptomeningeal and pial veins resulting in ischemic and haemorrhagic necrosis of the corresponding white matter and cortex. In these areas there was an increase of blood vessels

*Table 7*: Vascular events before diagnosis of homocystinuria, vascular abnormalities at physical and Doppler-flow examination at diagnosis of homocystinuria and after start of therapy, in 21 homozygous homocystinuria patients

| Patients | Before diagnosis | At diagnosis of homocystinuria | | | | | After start of therapy | | | |
|---|---|---|---|---|---|---|---|---|---|---|
| | events | year(age) | physical abnormalities | Doppler flow-abnormalities* | angiography performed** | events | start | physical abnormalities | Doppler flow-abnormalities | events |
| no 1 | pulmonary embolisms, 1969 | 1979(50) | - | carotid aa[&]<br>vertebral aa | - | - | 1979 | - | -<br>- | - |
| no 2 | - | 1978(20) | - | carotid aa | - | - | 1978 | - | - | - |
| no 3 | migraine, 1974 | 1978(20) | aortic valve incompetence | brachial aa | - | - | 1978 | = | - | - |
| no 4 | - | 1978(25) | - | vertebral a | - | - | 1979 | - | - | - |
| no 5 | recurrent thrombophlebitis, 1975<br>deep venous crural thrombosis, 1975 | 1978(27) | pulseless dorsal pedis a | carotid aa | - | - | 1978 | = | - | - |
| no 6 | - | 1977(21) | hypertension<br>ischemic left toes; pulseless dorsal pedis a;<br>abdominal bruit | femoral a | + | - | 1977 | =<br>-<br>=<br>- | - | - |
| no 7 | carotid aa-stenosis, 1975<br>hypertension, 1975 | 1977(29) | hypertension | carotid a | + | - | 1978 | = | ↓1983;-1984[b] | - |
| no 8 | - | 1968(22) | hemiplegia | ? | - | cerebrovascular accident; calf vein thrombosis | 1968 | hypertension, 1980 | vertebral a 1980;-1983 | - |

| | | | | | | | | | | |
|---|---|---|---|---|---|---|---|---|---|---|
| no 9 | severe migraine, 1978 | 1980(20) | hemiplegia papiledema | ? | + | died from cerebro-vascular accident | | | | |
| no 10 | - | 1970( 4) | hypertension | ? | - | - | 1970 | - | -1982;= | - |
| no 11 | - | 1970( 4) | - | ? | - | - | 1971 | - | -1982;= | - |
| no 12 | recurrent cerebro-vascular accidents, 1966; bulbar palsy, 1973 carotid aa-stenosis, 1978 | 1981(54) | hemiplegia bulbar palsy; | carotid aa brachial aa | + | myocardial infarction | 1981 | = | ? | - |
| no 13 | deep venous crural thrombosis, 1949 | 1982(53) | - | - | | - | 1982 | - | - | - |
| no 14 | angina pectoris, 1956 | 1982(50) | heart murmur | ? | + | - | 1984 | = | -1984 | - |
| no 15 | - | 1981(31) | - | carotid a | - | - | no treatment | | | |
| no 16 | - | 1982(15) | mitral valve prolapse | ? | - | - | 1982 | = | -1984 | - |
| no 17 | - | 1982(11) | ? | ? | - | - | 1982 | ? | ? | ? |
| no 18 | - | 1984(23) | - | carotid a | - | - | 1984 | | | |
| no 19 | - | 1984( 9) | - | - | - | - | 1984 | | | |
| no 20 | crural thrombophlebitis, 1984 | 1984(46) | - | - | - | - | 1984 | | | |
| no 21 | - | 1975( 9) | deep venous crural thrombosis | ? | - | peptic ulcer | 1975 | - | ? | - |

* decreased velocity of blood flow at a local stenosis or increased diastolic velocity as a sign of systemic arteriosclerosis

** results of angiography performed in 5 patients are presented in the text

& aa = arteries; a = artery

þ result of Doppler flow-examination less abnormal in 1983 compared to the result at diagnosis; normal result in 1984

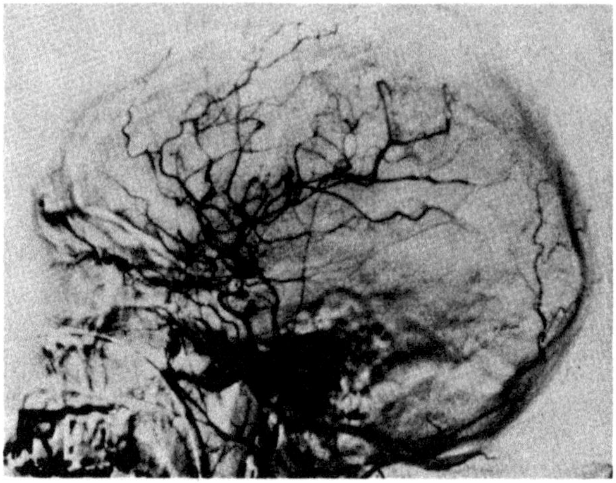

*Fig. 5.* Lateral projection of right carotid angiogram in patient no 9; note extreme narrowing of the extra and intracranial vessels with elevation of the lateral part of the middle cerebral artery and of the Sylvian triangle.

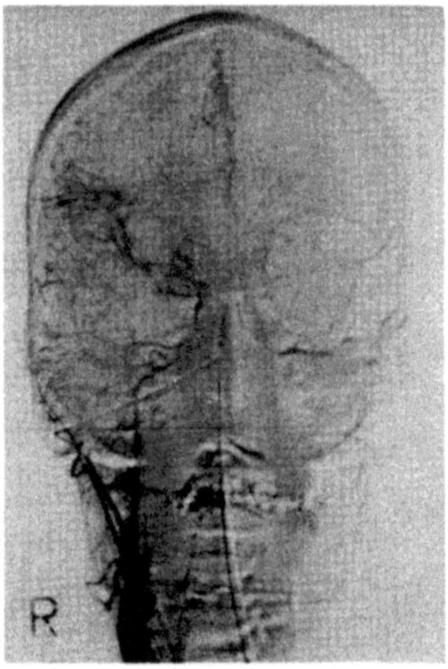


*Fig. 6.* Posterior-anterior projection of right carotid angiogram in patient no 9; note that the anterior cerebral artery is displaced towards the left side.

and swelling of the endothelial cells.In spite of the patient's young age, a remarkable number of hyaline-thickened vessels was present, sometimes with splitting of the underlying internal elastic lamina, especially in the middle cerebral artery. These findings are well in keeping with endothelial cell damage and complicating thromboembolism in homocystinuria patients (Chapter 1.7.2).

*Patient no 12*, a female born in 1927, had suffered since 1966 from recurring cerebrovascular accidents resulting in hemiplegia. In 1973 bulbar palsy was established and in 1974 the pulsations of the left radial artery were found to be absent. Angiography performed in 1978 demonstrated complete stenosis of the right internal carotid artery and partial stenosis of the left one. In 1981 she was admitted because of pneumonia. Further evaluation disclosed a recent myocardial infarction and stenoses of the carotid and the brachial arteries at Doppler flow examination. Re-evaluation of vascular pathology revealed, apart from the occlusion of the right carotid artery and an atherosclerotic ulcer in the left carotid artery, multiple irregularities of the walls of all cerebral arteries and the left subclavian artery. These striking arteriosclerotic lesions and thromboembolic events in this young patient led to the investigation for the presence of homocystinuria.The levels of amino acids in fasting serum, however, were normal (Chapter 2.3). Nonetheless, homozygous homocystinuria could be established on the strength of an abnormal rise of total homocysteine in the blood after methionine loading in combination with deficient cystathionine synthase activity in her cultured fibroblasts (Chapter 2.4). After the start of pyridoxine in combination with acenocoumarol in 1981, no further vascular events occurred in this patient but physical and mental deterioration progressed.

In *patient no 14* homocystinuria was diagnosed at the age of 50 when family screening was performed among sibs of her sister, patient no 13. From the age of 24 she suffered from angina pectoris. Coronary angiography at the age of 49 revealed arteriosclerosis. There was a heart murmur without signs of valve involvements at echocardiographical examination. She refused therapy with pyridoxine until 2 years after diagnosing homocystinuria and she rejects anticoagulant treatment up to the present.

*Patient no 21* suffered at the age of 9 from peptic duodenal ulcer and deep venous crural thrombosis. These events and the earlier established ectopia lentis, oligophrenia, and epilepsy led to the investigation for the presence of homocystinuria. Neither angiographic nor Doppler flow examinations have been performed. After the start with pyridoxine in 1975 no further vascular events happened.

In another 7 patients vascular involvement had led to minor events and abnormalities, such as recurring pulmonary embolisms in patient no 1 during longstanding periods of bed rest because of her psychiatric illness, deep venous crural thrombosis in patient no 5 after minor knee surgery and in patient no 13 in the puerperium, and periods of crural thrombophlebitis in patient no 20. In patient no 10 hypertension (150/90 mmHg) has been established at the time of the diagnosis at the age of 4, a mitral valve prolapse appeared to be present in patient no 16, and patient no 3 suffered from migraine.

After the start of treatment for homocystinuria vascular abnormalities remained absent at physical examination in the patients nos 1 and 13 and were unchanged in the patients nos 3, 5, 16 and 20. High blood pressure in patient no 10 normalized. Abnormal Doppler flow characteristics were established in the patients nos 1, 3 and 5 at diagnosis of homocystinuria and these became normal in the follow-up periods ranging from 5 to 6 years after the start of treatment. No abnormal Doppler flow characteristics have been found in the patients nos 10 and 13 at first examination in 1982 and at yearly follow-up until 1984. In the patients nos 16 and 20 such studies were firstly performed in 1984 and no abnormalities have been established in them. In none of these 7 patients vascular events occurred after the start of treatment for periods ranging from ½ to 14 years.

In the remaining 7 patients, vascular pathology has been established neither clinically nor in their morbid history. Notwithstanding that, in 4 of them Doppler flow examination clearly showed abnormal characteristics, in patients nos 2, 15 and 18 of one or both carotid arteries, and in patient no 4 of one vertebral artery. Such examination did not reveal any abnormality in patient no 19 at age of 9 years at the diagnosis of homocystinuria and in patient no 11 in whom this non-invasive investigation was peformed as late as 12 years after making the diagnosis of homocystinuria. Patient no 17 has not been studied by this technique. After the start of treatment in these patients no vascular abnormalities have been established at physical examination and, on follow-up, Doppler flow abnormalities disappeared in the patients nos 2 and 4. Patient no 15 has not been examined after establishing the diagnosis, and patient no 18 started with treatment at the end of 1984. In none of these patients vascular events occurred after the start of treatment for periods ranging from ½ to 11 years.

In their survey of more than 600 homocystinuria patients, Mudd et al[3] reported a statistically significant reduction of the number of initial and recurrent thromboembolic events in pyridoxine responsive homocystinurics while on treatment (Chapter 1.11). From the data in our relatively small number of patients a similar conclusion is not justified. Nevertheless, the absence of vascular complications in all patients after the start of treatment and the normalization of Doppler flow characteristics in the patients who showed abnormalities at the time of the diagnosis suggest some beneficial clinical effect. It has to be admitted, however, that the uneventful clinical outcome up the present in this group of homocystinuria patients after the start of homocysteine-lowering treatment may be in part the result of the simultaneously prescribed anticoagulant therapy with acenocoumarol in about 50% of them.

Finally, it has to be stressed that invasive angiographic examinations should be used with due caution in homocystinuria patients[6,20]. Even the minor trauma of puncture with a catheter in the arterial and venous wall might lead to thromboembolic events. In patient no 6 severe abdominal pain developed within hours after abdominal angiography. Ischemia of the colon was held to be its cause. During the following days the pain subsided gradually and the patient recovered completely. In patient no 9 angiography has been performed after the development of clinical signs of a cerebrovascular accident with brain edema. It cannot be completely excluded that the multiple fresh cerebral thromboses in addition to the expected temporal infarction established at autopsy have been luxated by the performed angiographic manoeuvre[6].

### *2.6.4 Neurologic and psychiatric disorders (Table 8)*[21]

IQ values at the time of the diagnosis of homocystinuria are available only in 8 out of the 21 homozygous patients, whereas the remaining 13 patients were rated as being of average or low-average intelligence, or mildly or grossly mentally retarded. Seven out of the 21 patients appeared to be of average and 9 of low-average intelligence, whereas 3 were mildly and 2 grossly retarded. In the patients nos 16, 19 and 21 psychomotor retardation has been established 13, 2 and 7 years before the diagnosis of homocystinuria respectively, and 2 of them proved to be grossly retarded at the time of the diagnosis. It is noteworthy that the only pyridoxine non-responsive patient in this group of homocystinurics also appeared to be the most severely retarded one (Chapter 1.6.4). Two of the "partial" pyridoxine responsive patients, nos 10 and 11, have been treated from very young ages on when IQ values of 100 were measured. Up to the present, 14 respectively 13 years after the start of treatment, their intelligence is still rated as being average. In patient no 19 such follow-up evaluation after the start of therapy is not yet possible.
It has to be stressed that from these data as well as from the recently reported figures by Mudd et al[3], the conclusion seems justified that mental retardation, at least in pyridoxine responsive homocystinuria patients, is not at all a consistent finding. In our group of 20 pyridoxine responsive patients only 4 (20%) could be classified as mildly or grossly retarded, whereas Mudd et al. established such retardation in 38% in their much larger number of responsive patients. Therefore, it is completely unreliable to differentiate between homocystinuria and e.g. Marfan's syndrome merely on the strength of the absence of mental retardation in the latter (Chapter 4).

Seizures occurred in no more than 4 (19%) of the homozygous homocystinuria patients, in patient no 5 for a short period at the age of 4 years and in the patients nos 8 and 9 seizures developed after the event of a cerebrovascular accident. In patient no 21 epilepsy developed at the age of 3, unrelated to a vascular accident, 6 years before establishing homocystinuria. Electroencephalographic abnormalities, however, have been observed in 15 (75%) of the

*Table 8*: Neurologic and psychiatric disorders before and at the time of the diagnosis of homocystinuria, and after the start of therapy, in 21 homozygous homocystinuria patients

| Patients | Before diagnosis | At diagnosis of homocystinuria | | | | | After start of therapy | | | |
|---|---|---|---|---|---|---|---|---|---|---|
| | events | year(age) | IQ⁺ | seizures | electroencephalographic abnormalities | psychiatric illness | start | seizures | electroencephalographic abnormalities | psychiatric illness |
| no 1 | depressive and paranoid psychosis 1969-1974 | 1979(50) | 86 | - | slight and aspecific | depression alcoholism | 1979 | - | $\downarrow$1983 | $\downarrow$1984 -1984 |
| no 2 | - | 1978(20) | low-average | - | slight and aspecific | - | 1978 | - | ->1980 | - |
| no 3 | - | 1978(20) | 82 | - | slight and aspecific | - | 1978 | - | ->1984 | - |
| no 4 | - | 1978(25) | average | - | - | - | 1979 | - | - | - |
| no 5 | seizures 1955 | 1978(27) | low-average | - | - | - | 1978 | - | - | - |
| no 6 | - | 1977(21) | 94 | - | slight and aspecific | - | 1977 | - | ->1982 | - |
| no 7 | paranoid depression and suicidal tendencies 1975 | 1977(29) | 90 | - | aspecific | depression | 1978 | - | sharp waves >1980;->1983 | - |
| no 8 | - | 1968(22) | mildly retarded | + | sharp and slow waves | - | 1968 | +, ->1975 | ->1983 | - |
| no 9 | - | 1960(20) | average | + | epileptic state | - | deceased in 1980 | | | |
| no 10 | - | 1970( 4) | 100 | - | slow waves | - | 1970 | - | ->1982 | - |
| no 11 | - | 1970( 3) | 100 | - | slow waves | - | 1971 | - | ->1984 | - |

| | | | | | | | | | | |
|---|---|---|---|---|---|---|---|---|---|---|
| no 12 | depression and suicidal tendencies 1974 | 1981(54) | average | - | aspecific | - | 1981 | - | ? | - |
| no 13 | - | 1982(53) | average | - | - | - | 1982 | - | slight and aspecific >1983 | - |
| no 14 | puerperal psychosis 1956-1959 | 1982(50) | low-average | - | - | - | 1984 | | | |
| no 15 | - | 1981(31) | average | - | aspecific | - | no treatment | | | |
| no 16 | psychomotor retardation 1969 oligophrenia 1973 | 1982(15) | 41 | | aspecific | - | 1982 | - | = | - |
| no 17 | - | 1982(11) | mildly retarded | - | ? | - | 1982 | ? | ? | ? |
| no 18 | - | 1984(23) | low-average | - | - | - | 1984 | | | |
| no 19 | psychomotor retardation 1982 | 1984( 9) | 92 | - | slight and aspecific | - | 1984 | | | |
| no 20 | depression 1972 | 1984(46) | low-average | - | aspecific | depressive psychosis | 1984 | | | |
| no 21 | psychomotor retardation 1968 seizures 1969 | 1975( 9) | grossly retarded | + | sharp and slow waves | - | 1975 | ↑>1980 | ↑>1980 | - |

[+]see text

20 patients in whom such examination has been performed at the time of the diagnosis of homocystinuria. In 10 of them the abnormalities were aspecific and only minor in most of the cases. In the patients nos 10 and 11 an abnormal pattern of slow wave-activity (Chapter 1.6.4) has been established, and in the patients nos 8, 9 and 21 also sharp wave-discharges in accordance with their epileptic manifestations were observed.

After the start of treatment seizures did not occur up to the present in those 18 patients who did not show this abnormality at the time of the diagnosis of homocystinuria. In patient no 8 epilepsy did not happen since 1975, 7 years after the start with pyridoxine. Patient no 9 died at the time of the diagnosis from a cerebrovascular accident complicated by coma, an epileptic state, and apnea (Chapter 4). Only in patient no 21 there was a progress of the frequency and severeness of the seizures since 5 years after the start of treatment with pyridoxine. Of the 13 patients in whom electroencephalographic characteristics could be evaluated at yearly examinations since starting therapy, 7 showed a normalization of their abnormal signs after one or more years. In one patient the characteristics remained unchanged, whereas in another one the abnormality decreased. In patient no 21 the epileptic electroencephalographic signs were progressive despite treatment since 5 years. In 2 of the 3 patients who did not show abnormalities at the time of diagnosis of homocystinuria, abnormal signs did not appear after the start of treatment either. Only in patient no 13 electroencephalography became abnormal.

Psychiatric illnesses occurred in 5 (24%) of the 21 patients before or at the time of the diagnosis of homocystinuria. In these patients severe depression was present, in the patients nos 14 and 20 also with psychotic characteristics. In the patients nos 7 and 12 the depression had led to attempted suicide. Among the sibs of these 5 patients no psychiatric disorders have been established as yet. From these data psychiatric illness seems more likely to be a symptom of the homocystinuria syndrome, than a mere chance association, as has been suggested in literature (Chapter 1.6.4).

In 2 of 3 patients presenting psychiatric illness at the time of the diagnosis of homocystinuria, the disorder resolved soon after the start of treatment with pyridoxine. Since then these patients are able to live a completely normal life. In patient no 1 the depressive periods became less severe and long-term institutionalization as needed before the diagnosis of homocystinuria, was no longer required. She was cured of alcohol abuse by successful hypnotherapy. In none of the other patients psychiatric illnesses occurred since the start of treatment.

### *2.6.5 Miscellaneous (Table 9)*

A fair, thin skin (Chapter 1.6.5) and livedo reticularis (Chapter 1.6.3) were observed in 40% of the homocystinuria patients; in half of them both signs were present. Malar flush (Chapter 1.6.3) demonstrated no more than 15% of these homocystinurics. Thirty percent of the patients had blond, brittle hair and 50% blue and light irides (Chapter 1.6.5). A marked increase of the incidence of inguinal hernia is not indicated from our data as only in patient no 18 such hernia occurred; in the patient no 11 an umbilical as well as an epigastrical hernia was established and this patient also developed torsio testis at the age of 15.

Several events and abnormalities seemingly unrelated to the homocystinuria syndrome have been observed in these patients (Table 9). The patients nos 4 and 5, two brothers, showed renal glucosuria, whereas the latter also developed severe Menière's disease at the age of 33. In patient no 7 two malignancies occurred, an ovarial cystadenocarcinoma at age of 19 and malignant melanoma of the skin of the leg at the age of 34. No signs of metastases have been established as yet. In patient no 21 a peptic duodenal ulcer occurred at the age of 9; although such event is not related to homocystinuria, it had led to its investigation in this patient who already showed ectopia lentis, mental retardation, and seizures at younger age.

## 2.7 OUTCOME OF PREGNANCIES IN 7 HOMOZYGOUS HOMOCYSTINURIA FEMALES (Table 10)

Up to January 1985 eight of the 21 homocystinuria patients were married, 7 female and one male patient. This male homocystinuric, patient no 4, fathered 3 normal children. The outcomes of the pregnancies in the 7 female homocystinurics are shown in Table 10. The only pregnancy in which the mother received therapy for her homocystinuria, was the second pregnancy in patient no 3. This pregnancy was uneventful and a normal child was born,

*Table 10*: Outcome of pregnancies in 7 female homozygous homocystinuria patients

| Patients | Age at January 1985 yr | Full term normal child | Spontaneous abortion | Immature birth dead child | Anticonception | age yr |
|---|---|---|---|---|---|---|
| no 1 | 55 | 3 | 1 | – | – | |
| no 3 | 26 | 1* | 1 | – | sterilization | 23 |
| no 7 | 36 | 2 | – | – | IUD | 25 |
| no 12 | 58 | 2 | 1 | – | – | |
| no 13 | 57 | 3 | – | 1 | – | |
| no 14 | 52 | 4 | 1 | – | sterilization | 37 |
| no 20 | 46 | 1** | 5 | 1 | – | |

* pregnancy in which the patient received therapy for homocystinuria
** child suffering from pseudohypoparathyroidism since the age of 11 years

*Table 9*: Miscellaneous findings in 21 homozygous homocystinuria patients

| Patients | Fair skin | Livedo reticularis | Malar flush | Fair, brittle hair | Blue or light iris | Events and abnormalities related to homocystinuria | unrelated |
|---|---|---|---|---|---|---|---|
| no 1 | + | + | - | + | + | hernia nuclei pulposi-surgery, 1955; varicosectomy, 1960 | tuberculosis of the lung and hip, 1955 |
| no 2 | + | + | - | + | + | - | - |
| no 3 | + | - | - | - | + | - | - |
| no 4 | - | - | - | - | + | - | renal glucosuria, 1978 |
| no 5 | - | + | - | - | - | - | renal glucosuria, 1978; Menière's disease, 1984 |
| no 6 | - | - | - | - | - | "rachitis", 1959 | - |
| no 7 | - | - | - | - | - | - | cystadenocarcinoma ovarii, 1968; malignant melanoma, 1983 |
| no 8 | - | + | - | - | - | - | hemistrumectomy because of benignant adenoma, 1980 |
| no 9 | + | - | - | + | - | - | - |
| no 10 | - | - | - | - | - | - | - |
| no 11 | - | - | - | - | + | umbilical hernia, 1970; epigastrical hernia, 1975; torsio testis, 1982 | - |

| | | | | | | | |
|---|---|---|---|---|---|---|---|
| no 12 | - | - | - | - | + | - | - |
| no 13 | - | + | + | - | - | - | peptic duodenal ulcers, 1980-1984 |
| no 14 | + | - | - | + | + | "encephalitis", 1933 | recurrent nephrolithiasis, 1968-1984 |
| no 15 | - | - | - | - | - | - | - |
| no 16 | + | + | - | - | - | - | - |
| no 17 | ? | ? | ? | ? | ? | ? | ? |
| no 18 | - | - | - | - | + | inguinal hernia surgery, 1962 | - |
| no 19 | + | + | + | + | + | - | - |
| no 20 | - | - | + | - | - | - | - |
| no 21 | + | + | - | + | + | - | peptic duodenual ulcer, 1975 |
| | 8+<br>12– | 8+<br>12– | 3+<br>17– | 6+<br>14– | 10+<br>10– | | |

in contrast to the first pregnancy which ended with a spontaneous abortion. The table shows that 26 pregnancies occurred in the 7 patients when untreated with an outcome of spontaneous abortion in 9 pregnancies and immature birth of a stillborn child in 2. This high rate of fetal loss (42%) is far beyond that in the general population living in the rural area around this hospital (15-20%)[22]. However, this result is strongly influenced by the major contribution from patient no 20 who had 5 spontaneous abortions and one stillbirth. Exclusion of the data from this patient would lead to 19 pregnancies in 6 homocystinurics mothers with fetal loss in 26%, a rate not markedly exceeding that in the general population.

Pregnancy is held to be an additional risk of thromboembolic events in homocystinurics (Chapter 1.10.3). In the 27 pregnancies in these 7 patients however, not such event occurred except for a deep venous crural thrombosis in patient no 13 in the puerperium. In patient no 14 a psychosis developed in the puerperium of her first pregnancy, from which she suffered for as long as 3 years.

It is well-known that fetuses of phenylketonuric mothers may develop congenital abnormalities. In contrast, no such abnormalities reportedly occur in fetuses of homocystinuric mothers[3]. The 16 children delivered from the female patients from our group of homocystinurics all were normal at birth and do not suffer from any illness at ages at the present time ranging from 4 to 38 years, except for the child of patient no 20 in whom pseudohypoparathyroidism was established at age of 11 years. It is clear that all these children are obligate heterozygotes for homocystinuria due to cystathionine synthase deficiency. Data on amino acid levels in the fasting state and after a standardized methionine load and determinations of cystathionine synthase activity in cultured skin fibroblasts in 8 of them are presented in Chapter 7, where these characteristics obtained in a total of 20 obligate heterozygotes are discussed.

## 2.8 FAMILY SCREENING (Table 11)

Family screening was conducted in 15 of the 16 families (Table 11). Only in the sibships nos 4 and 6 the parents were consanguineous. The number of further sibs in these 15 sibships was 49 and in 36 of these subjects screening for homocystinuria has been performed. In the sibs of sibships nos 1 and 2 only quantitative amino acid analysis of urine was available, which method – in our hands – does not differentiate between heterozygotes for homocystinuria and normal subjects. In all other sibs a standardized methionine loading test has been performed, and, moreover, determination of cystathionine synthase activity in cultured fibroblasts from a skin biopsy in those subjects whose levels of total homocysteine after methionine loading exceeded the ranges established in control subjects. As will be discussed in Chapter 7, the combi-

*Table 11*: Family screening among the sibs of 20 homozygous homocystinuria patients

| Patients | Sibships | Further sibs in sibships | | | | |
|---|---|---|---|---|---|---|
| | | Number | Homozygote | Heterozygote | Normal | Unknown |
| no 1 | 1 | 7 | | ? | 6* | 1** |
| no 2 | 2 | 4 | | ? | 4* | |
| no 3 | 3 | 5 | pat. no 4<br>pat. no 5 | 1 | 2 | |
| no 6 | 4 | 5 | pat. no 7 | 4 | | |
| no 8 | 5 | 2 | | | | 2⁰ |
| no 9 | 6 | 2 | | 2 | | |
| no 10 | 7 | 1 | pat. no 11 | | | |
| no 12 | 8 | 3 | | | | 3 |
| no 13 | 9 | 6 | pat. no 14 | | 1 | 4 |
| no 15 | 10 | 2 | | 1 | 1 | |
| no 16 | 11 | 2 | | 2 | | |
| no 17 | 12 | ? | | | | |
| no 18 | 13 | 1 | | | 1 | |
| no 19 | 14 | 1 | | 1 | | |
| no 20 | 15 | 7 | | 2 | 2 | 3& |
| no 21 | 16 | 1 | | 1 | | |
| Total | | 49 | 5 | 14 | 17 | 13 |

* examined only by quantitative amino acid analysis of urine, not detecting heterozygotes
** died at age of 35 years because of cerebrovascular accident
⁰ one of these sibs died at age of 25 years because of pulmonary embolism
& one of these sibs died at age of 5 months from unknown cause

nation of the results of both these tests discriminated all carriers out of a group of 20 obligate heterozygotes for homocystinuria from normal control subjects without exception.

Table 11 shows that homozygosity for homocystinuria has been established in 5 out of the 36 sibs who were screened for this disorder, heterozygosity in 14 and a normal condition in 17 of these sibs. After classifying the 13 sibs with unknown status according to the chance rate of 1: 2: 1 for being homozygote, heterozygote, and normal, and assuming that out of the 10 "normal" sibs from the sibships nos 1 and 2 at least 6 persons should be heterozygote, the ratio between homozygotes, heterozygotes, and normal subjects then might be 8: 27: 14 in these 49 further sibs, whereas the expected ratio is 12: 25: 12 (chi square test $P > 0.10$).

It is intriguing that out of the group of 13 sibs with unknown genetic status, 3 sibs died at very early age, 2 of them from causes suggestive of the presence of homocystinuria.

SUMMARY

This group of 21 studied homozygous homocystinuria patients is rather unique because in 15 of them the diagnosis of homocystinuria has been established at their adult age, after a considerable delay between the appearance of the first symptoms of the disorder and the date of making the diagnosis of homocystinuria. All of these late detected homocystinurics appeared to be pyridoxine responsive and it is known from literature that suchlike patients suffer from fewer and more mild symptoms of the homocystinuria syndrome than non-responders do. Indeed, the only pyridoxine non-responsive patient out of this group of homocystinurics (no 16) was also the most severely affected one. Among the pyridoxine responsive patients there was a large heterogeneity of the clinical expression of symptoms. One patient (no 4) does not show any clinical abnormality up to the present at the age of 32. Most patients are mildly affected (nos 2, 3, 5, 10, 11, 13, 15, 17, 18 and 19), showing ectopia lentis as their main feature. However, several patients (nos 1, 6, 7, 8, 9, 12, 14, 16, 20 and 21) have in addition presented major complications or abnormalities, in particular vascular events (nos 6, 7, 8, 9, 12 and 14), severe scoliosis and oligophrenia (nos 16 and 21), or serious psychiatric illnesses (nos 1, 7, 12 and 20).

The most consistent clinical findings among the 21 homocystinuria patients in this group were scoliosis (81%), ectopia lentis with myopia (76%), and marfanoid features (62%). Moreover, further evaluation disclosed involvement of the vascular system in 18 patients (86%), leading to major vascular complications in 7 of them. One patient (no 9) died at the age of 20 from a cerebrovascular accident complicated by coma, an epileptic state, and apnea. Furthermore, osteoporosis has been established in 14 patients (82%), and electroencephalographic abnormalities in 75% of the patients, albeit only minor and aspecific abnormal characteristics in most of them.

After the start of therapy for homocystinuria either with pyridoxine in high dose as the sole treatment, or in combination with folic acid, betaine, with or without a methionine-restricted diet, the clinical outcome has been favourable in all patients, except in one (no 21) in whom the frequency of seizures, the scoliosis, and the mental deterioration have proven to be progressive. In another patient (no 3) retinal detachment occurred 6 years after the start with pyridoxine, but this complication is probably more due to the surgical lens extraction 11 years before establishing the diagnosis of homocystinuria, than to the mere presence of homocystinuria[17].

It has to be stressed that in 2 patients (nos 12 and 15) the diagnosis of homozygous homocystinuria could not be established on the strength of an abnormal profile of amino acids in fasting serum. Homozygosity in these patients was unmasked only after performing an oral methionine loading test and determination of cystathionine synthase activity in their cultured fibroblasts.

## REFERENCES


1. Drayer JIM, Cleophas AJM, Trijbels JMF, Smals AGH, Kloppenborg PWC. Symptoms, diagnostic pitfalls, and treatment of homocystinuria in seven adult patients. Neth J Med 1980; 23: 89–94.
2. Monnens L, Trijbels F, van der Zee S. Homocystinurie: ervaringen met de behandeling. Maandschr Kindergeneesk 1972; 40: 240–53.
3. Mudd SH, Skovby F, Levy HL, Pettigrew KD, Wilcken B, Pyeritz RE, Andria G, Boers GHJ, Bromberg IL, Cerone R, Fowler B, Gröbe H, Schmidt H, Schweitzer L. The natural history of homocystinuria due to cystathionine β-synthase deficiency. Am J Hum Genet 1985; 37: 1–31.
4. Cruysberg JRM, Boers GHJ, Pinckers A, Schoonderwaldt HC. Homocystinuria, a frequently missed diagnosis? Ophthalmologica, Basel 1984; 188: 189–90.
5. Boers GHJ, Polder TW, Cruysberg JRM, Schoonderwaldt HC, Peetoom JJ, van Ruyven TWJ, Smals AGH, Kloppenborg PWC. Homocystinuria versus Marfan's syndrome: the therapeutic relevance of the differential diagnosis. Neth J Med 1984; 27: 206–12.
6. Schoonderwaldt HC, Boers GHJ, Cruysberg JRM, Schulte BPM, Slooff JL, Thijssen HOM. Neurologic manifestations of homocystinuria. Clin Neurol Neurosurg 1981; 83: 153–62.
7. Spaeth GL, Barber GW. Prevalence of homocystinuria among the mentally retarded: evaluation of a specific test. Pediatrics 1967; 40: 586–9.
8. Spaeth GL. Screening for homocystinuria. Survey Ophthalmol 1984; 29: 230.
9. Brenton DP, Cusworth DC. The response of patients with cystathionine synthase deficiency to pyridoxine, In: Carson NAJ, Raine DN, eds. Inherited disorders of sulphur metabolism. Edinburgh: Churchill Livingstone, 1971: 264–74.
10. Fowler B, Sardharwalla IB. Homocystinuria: cystathionine synthase activity in cultured skin fibroblasts. In: International Symposium on inborn errors of metabolism in humans. Interlaken, 1980, 20.
11. Boers GHJ, Fowler B, Smals AGH, Trijbels JMF, Leermakers AI, Kleijer WJ, Kloppenborg PWC. Improved identification of heterozygotes for homocystinuria due to cystathionine synthase deficiency by the combination of methionine loading and enzyme determination in cultured fibroblasts. Hum Genet 1985; 69: 164–9.
12. Boers GH, Smals AG, Trijbels FJ, Leermakers AI, Kloppenborg PW. Unique efficiency of methionine metabolism in premenopausal women may protect against vascular disease in the reproductive years. J Clin Invest 1983; 72: 1971–6.
13. Mudd SH, Levy HL. Disorders of transsulphuration. In: Stanbury JB, Wyngaarden JB, Fredrickson DS, eds. The metabolic basis of inherited disease. New York: McGraw-Hill Book Co, ed. 5, 1983: 522–59.
14. Carson NAJ, Carré IJ. Treatment of homocystinuria with pyridoxine. A preliminary study. Arch Dis Child 1969; 44: 387–92.
15. Gaull GE, Schaffner F. Electron microscopic changes in hepatocytes of patients with homocystinuria. Pediatr Res 1971; 5: 23–32.
16. Kang ES, Ryers RK, Gerald PS. Homocystinuria. Response to pyridoxine. Neurology 1970; 20: 503–7.
17. Cruysberg JRM, Boers GHJ, Pinckers A. Ocular abnormalities in homocystinuria. (in preparation).
18. Rahman M. Homocystinuria. Review of four cases. Br J Ophthalmol 1971; 55: 338–42.
19. Parish JG. Radiographic measurements of skeletal structure of the normal hand. Br J Radiol 1966; 39: 52–62.
20. Wicherink-Bol HF, Boers GHJ, Drayer JIM, Rosenbusch G. Angiographic findings in homocystinuria. Cardiovasc Intervent Radiol 1983; 6: 125–8.

21. Schoonderwaldt HC, Boers GHJ Colon E. Neurologic and psychiatric abnormalities and their evaluation after the start of therapy for homocystinuria, in a group of homozygous homocystinurics. (in preparation).
22. Annual report 1983. Department of Obstetrics and Gynaecology, St. Radboud Hospital, University of Nijmegen, Nijmegen, The Netherlands.

*Chapter 3*

# Pyridoxine treatment does not prevent homocystinemia after methionine loading in adult homocystinuria patients

Godfried H.J. Boers, Anthony G.H. Smals, Jan I.M. Drayer, Frans J.M. Trijbels, Annelies I. Leermakers and Peter W. Kloppenborg

*Division of Endocrinology, Department of Medicine and Department of Pediatrics, University of Nijmegen, Nijmegen, The Netherlands.*

## ABSTRACT

Fasting homocystinemia in homocystinuria due to cystathionine synthase deficiency reportedly disappears on high-dose pyridoxine treatment. This does not necessarily reflect normal tolerance to methionine. The present study compares the effects of oral methionine loading on homocystine, cystine, and homocysteine-cysteine disulphide profiles in 8 adult homozygous homocystinuria patients on and off pyridoxine treatment and in 20 controls. Pyridoxine nearly normalized fasting serum amino acid levels. Nevertheless, with a similar methionine load the patient's homocystine levels on and off treatment rose and the cystine levels decreased, reflecting the ongoing formation of the homocysteine-cysteine disulphide in the presence of impaired transsulphuration of homocysteine. In the controls homocystinemia remained virtually absent and cystine transiently rose which indicates normal transsulphuration. On treatment methionine loading evoked a brisk rise of the homocysteine-cysteine disulphide levels to values equal to those off treatment, when these levels virtually plateaued after the load. Thus, pyridoxine treatment attenuates the biochemical abnormalities in the fasting patients but leaves their impaired capacity to handle major methionine loads essentially unchanged.

## INTRODUCTION

Classical homocystinuria is an inborn, autosomal recessive error of methionine metabolism due to cystathionine synthase deficiency. Four major systems may be clinically involved: myopia and ectopia of the eye lens, skeletal abnormalities as osteoporosis, scoliosis and marfanoid features, premature and progressive vascular disease, and damage of the central nervous system.[1] Cystathionine synthase is the rate limiting enzyme in the transsulphuration of homocysteine to cysteine. Pyridoxal phosphate acts as cofactor in this reaction. In the clinical laboratory the combination of elevated levels of circula-

*Metabolism 1983; 32:390-7.*

ting homocystine and its precursor methionine, decreased concentrations of cystine and increased formation of homocysteine-cysteine mixed disulphide in blood and urine is pathognomonic. These biochemical abnormalities can disappear after treatment for several weeks with high doses of pyridoxine and therefore homocystinuria due to cystathionine synthase deficiency is considered as a vitamin responsive disorder.[2-11]

Experimental studies suggest that recurrent arterial and venous thrombosis is the base of the widespread organ damage.[12,13] Accumulation of homocysteine in the endothelial cells might be one of the initiating events in these thrombogenetic processes.[14,15] If pyridoxine treatment could prevent this endothelial damage, such therapy could be clinically beneficial. The disappearance of homocystinemia by such treatment might be a useful biochemical indicator of the effectiveness of the therapy. Methionine in the food is the natural source of the accumulation of homocysteine in this disorder. Methionine loading normally does not reportedly lead to detectable levels of its demethylation product homocysteine in the blood and urine due to rapid transsulphuration of the latter to cysteine. In homocystinuria, however, accumulation of homocysteine after methionine loading can occur and scarce reports indicate that pyridoxine treatment cannot completely restore normal methionine metabolism. In 1978 Mudd et al. pointed out, however, that an empirical answer to this question is not provided yet.[16]

In the present study profiles of serum methionine, homocystine, cystine and the homocysteine-cysteine mixed disulphide after standardized oral methionine loading in 8 adult patients with homocystinuria due to proven cystathionine synthase deficiency on and off high-dose pyridoxine treatment are compared to those in 20 healthy volunteers. The results clearly indicate that vitamin $B_6$ treatment at least in these adult homozygous patients with almost total cystathionine synthase deficiency is unable to normalize metabolism of methionine.

## MATERIALS AND METHODS

### *Patients and Subjects*

The clinical data of 8 homocystinuria patients who were diagnosed during their adult life (mean age 30 yr, range 20-50), are depicted in Table 1. The characteristics of 7 of these patients were reported extensively elsewhere,[17] and are briefly summarized herein.

*Patient 1* was suffering from ectopia lentis. Her history included psychiatric disorders (psychotic depression, paranoid psychosis) and pulmonary embolism. Further evaluation at the age of 50 yr, when the diagnosis homocystinuria was made, disclosed histopathological signs of osteoporosis and sclerosis of both carotid arteries (Doppler flow examination). Homocystinuria was not diagnosed in any of her living family members ($n = 11$).

*Patient 2* was examined because of ectopia lentis at the age of 20 yr. She showed marfanoid features (dolichostenomelia, arachnodactyly) and signs of osteoporosis both on radiological and histopathological examination. Doppler flow characteristics included signs of sclerosis of the carotid arteries. In none of the living family members ($n = 5$) a diagnosis of homocystinuria could be established.

*Patient 3* suffered from low back pain in the presence of scoliosis vertebrae before she was referred for further evaluation because of ectopia lentis at the age of 20 yr. There were arachnodactyly and histopathological signs of osteoporosis. Doppler flow examination disclosed signs of sclerosis of the brachial arteries. The living family members ($n = 7$) were screened and in 2 of them homocystinuria was diagnosed (patients 4 and 5).

*Patient 4*. a brother of patient 3, had an unremarkable history and showed no clinical signs of homocystinuria at the age of 25 yr when the diagnosis was made by family screening. Nevertheless, minor signs of scoliosis and osteoporosis were established on radiological and histopathological examinations.

*Patients 5*, a brother of patients 3 and 4, had a history of convulsions in his childhood, ectopia lentis, and repeated thrombophlebitis after venapunctures and knee surgery. The diagnosis homocystinuria was made at the age of 27 yr by family screening. He showed arachnodactyly and scoliosis, and signs of sclerosis of the carotid arteries on Doppler flow examination.

*Patient 6* was referred to our department for further evaluation at the age of 21 yr because of hypertension, and acrocyanosis of the third and fourth toe of the left foot. Her history was unremarkable, except for ectopia lentis. She showed marfanoid features (dolichostenomelia) and scoliosis and osteoporosis on radiological and histopathological examination. Arteriography disclosed severe stenosis of the splenic, iliocolic, and superior mesenteric artery, and irregularities of the vessel walls of the renal and coeliac arteries. A diagnosis of homocystinuria was made in this patient and in addition in one of her 7 living family members (patient 7).

*Patient 7*, a sister of patient 6, had a history of psychiatric illnesses (organic psychosyndrome, suicidal tendencies). Further evaluation at that time showed severe stenosis of the carotid arteries. At the age of 29 yr, when the diagnosis homocystinuria was made by family screening, extensive examination disclosed high grade myopia without ectopia lentis, radiological and histopathological signs of osteoporosis, and on repeated arteriography severe stenosis of the carotid arteries, and in addition partial stenosis of the splenic artery and an aneurysm in the spleen.

*Patient 8* had a history of ectopia lentis. At the age of 22 yr he suffered from cerebral thrombosis complicated by hemiplegia, severe convulsions, and deep crural thrombophlebitis. His marfanoid features (dolichostenomelia, arachnodactyly) were remarkable. The diagnosis homocystinuria was made. Further evaluation disclosed signs of osteoporosis and scoliosis. Doppler flow examination revealed complete stenosis of one vertebral artery.

*Table 1.* Clinical data from eight homocystinuria patients

| Clinical finding | Patient Number 1 | 2 | 3 | 4 | 5 | 6 | 7 | 8 |
|---|---|---|---|---|---|---|---|---|
| Sex | female | female | female | male | male | female | female | male |
| Age at diagnosis (years) | 50 | 20 | 20 | 25 | 27 | 21 | 29 | 22 |
| Symptoms | | | | | | | | |
| Ocular | | | | | | | | |
| Ectopia lentis | + | + | + | - | + | + | - | + |
| Myopia | + | + | + | - | + | + | + | + |
| Glaucoma | - | + | - | - | - | + | - | + |
| Skeletal | | | | | | | | |
| Dolichostenomelia* | - | + | - | - | - | + | - | + |
| Arachnodactyly | - | + | + | - | + | - | - | + |
| Histopathological signs of osteoporosis | + | + | + | + | - | + | + | + |
| Radiological signs of osteoporosis | - | + | - | + | - | + | + | + |
| Radiological signs of "fish-vertebrae" | - | + | + | + | + | + | + | + |
| Radiological signs of scoliosis | - | + | + | + | + | + | - | + |
| Vascular | | | | | | | | |
| Thromboembolism | + | - | - | - | - | + | - | + |
| Thrombophlebitis | - | - | - | - | + | - | - | + |
| Abnormal Doppler characteristics of arterial blood flow | + | + | + | + | + | + | + | + |
| Angiographically established arteriosclerosis | | | | | | + | + | |

| | | | | | | | | |
|---|---|---|---|---|---|---|---|---|
| Central Nervous System | | | | | | | | |
| Convulsions | - | - | - | - | + | - | - | + |
| Electroencephalographic abnormalities | + | + | + | - | - | + | + | + |
| Psychiatric illnesses | + | - | - | - | - | - | + | - |

*Dolichostenomelia: thin, excessively lengthened long bones (reduction of value of upper/lower segment).

The control group consisted of 10 healthy women and 10 healthy men. The mean age in the group of women was 27 yr (range 14-45) and of the group of men 35 yr (range 28-49) ($^+p>0.10$). The age difference between the whole group of normal subjects and the patients was not statistically significant ($^+p>0.10$).

After an overnight fast a load with L-methionine was administered orally in a dose of 0.1 g (0.7 mmole)/kg of body weight at 9.00 A.M. to the patients and the control subjects. During the test they stayed on the metabolic ward and consumed a methionine-free breakfast and luncheon. Informed consent was obtained from all patients and volunteers after approval of the protocol by the hospital Ethical Committee.

The methionine loading tests in the control group were performed while they were taking no medicaments, vitamin supplements, or oral contraceptives. In the homocystinuria patients the methionine loading tests were performed both during chronic therapy with pyridoxine, 750 mg/day, for at least 1 yr and after stopping this medication for 4 wk.

In 2 homocystinuria patients on chronic therapy with pyridoxine, 750 mg/day, measurements of serum amino acids were performed at several times during a day on regular hospital diet (80 g protein with 2.0 g methionine, 300 g carbohydrates, and 80 g fat, 2200 kcal). Breakfast was served at 8.15 A.M., dinner at noon, and luncheon at 5 P.M. Both patients had been taking a normal nonrestricted nonenriched diet during the weeks preceding the measurements.

### *Methods*

The determinations of cystathionine synthase activity in skin fibroblasts cultures were performed by Dr. B. Fowler (Willink Biochemical Genetics Unit, Pendlebury, Manchester, UK), using a method previously reported[18] with a slight modification. The results are shown in Table 2.

Methionine, homocystine, cystine and homocysteine-cysteine mixed disulphide in the serum were measured by ion-exchange chromatography (Biotronik LC 2000 amino acid analyzer) in blood samples taken at 9.00 A.M. immediately before methionine loading and 1, 2, 3, 4, 6 and 8 hr after loading. In the 2 patients mentioned before the amino acid concentrations were measured also at 1, 2, 5, 8, 9, 10, 11 and 12 A.M. and 3, 4, 5, 6, 9 and 11 P.M. during one day. The serum for amino acid analysis was deproteinized within 15 min after sampling. Trace amounts of homocystine and homocysteine-cysteine mixed disulphide lower than 0.1 μmole/L are denoted as not detectable (ND).

Because homocysteine and cysteine are rapidly oxidized to the disulphides homocystine, cystine and homocysteine-cysteine mixed disulphide during the procedure of blood sampling and chromatography, only these latter com-

*Table 2.* Determination of activity of cystathionine synthase (C.S.) in skin fibroblasts in eight homocystinuria patients before and after the addition of pyridoxal phosphate to the cell extracts

| | C.S. Activity (nmole/hr/mg cell protein) | |
|---|---|---|
| Patient Number | Before Pyridoxal Phosphate | After Pyridoxal Phosphate |
| 1 | 0.14 | 0.27 |
| 2 | 0.04 | 0.10 |
| 3 | 0.03 | 0.18 |
| 4 | 0.03 | 0.10 |
| 5 | 0.03 | 0.11 |
| 6 | 0.05 | 0.10 |
| 7 | N.D. | N.D. |
| 8 | 0.09 | 0.37 |
| Normal range | 3.6–25.7 | 4.9–32.1 |

Values are averages of at least two determinations. In these measurements the method previously reported[18] was used with one modification: the cystathionine formed during the incubation was separated by paper chromatography (solvent: isopropanol/formic acid/water, 80:6:20) with development times of at least 60 hours to ensure the separation of $^{14}C$-cystathionine from an additional radioactive compound formed in the assay.

pounds are measured by amino acid analysis and they indirectly reflect the concentrations of free homocysteine and cysteine respectively in the blood.[19]

Statistical analysis was performed using Wilcoxon's paired test ($p$), Wilcoxon's two sample test ($^{+}p$), and Friedman's nonparametric analysis of variance ($^{++}p$).

## RESULTS

### *Effects of methionine loading in control subjects*

The mean fasting serum methionine level in 10 control women (22 ± 2 μmole/L) was significantly lower ($^{+}p < 0.01$) than in 10 control men (33 ± 3 μmole/L (Fig. 1, Table 3). After methionine loading the serum levels did no longer differ between women and men. Peak levels at 1 hr ranged from 429 to 1487 μmole/L, mean 942 ± 77. Thereafter, the serum levels steadily and significantly ($^{++}p<0.001$) decreased and were still clearly elevated ($p<0.01$) up to 8 hr after loading.

Fasting serum homocystine levels in these healthy women and men were not detectable (17 subjects) or very low (0.1, 0.2, and 0.4 μmole/L in 3 healthy men). After methionine loading homocystine levels remained undetectable in 9 out of 10 women and 4 out of 10 men. In 1 of the women and in 6 men homocystine levels rose to maximally 1.6 μmole/L at 8 hr.

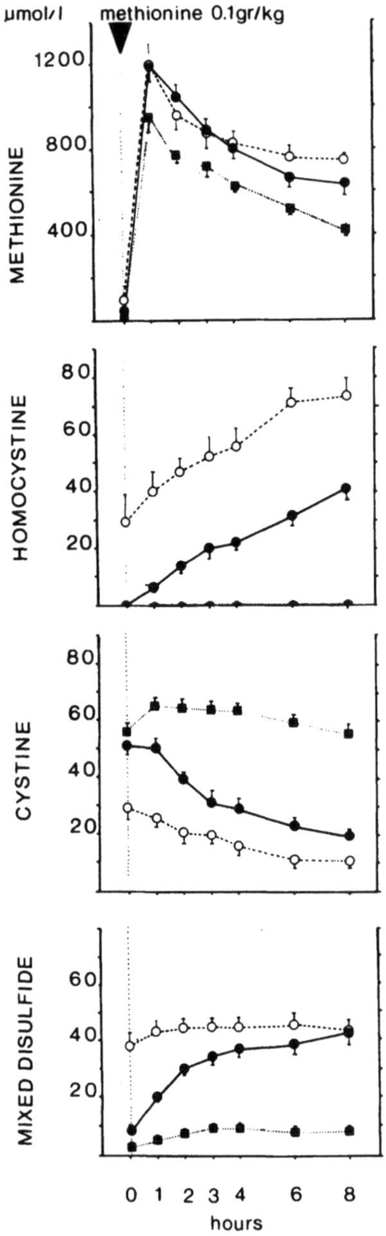


*Fig. 1* Mean concentrations (±SEM) of methionine, homocystine, cystine, and homocysteine-cysteine mixed disulphide in serum before and 1, 2, 3, 4, 6 and 8 hr after methionine loading (0.1 g/kg) in 20 normal subjects (■) and 8 patients with homocystinuria due to cystathionine synthase deficiency without pyridoxine treatment (○) and during chronic pyridoxine treatment, 750 mg/day (●).

The mean serum cystine levels did not differ between women and men (55 ± 2 and 58 ± 2 µmole/L respectively) neither in the fasting state nor after loading. After a significant albeit slight rise to 64 ± 2 µmole/L ($p<0.01$) at 1 hr after loading, the serum levels returned to preloading concentrations, 55 ± 2 µmole/L at 8 hr.

The mean fasting serum homocysteine-cysteine mixed disulphide levels in women (0.9 ± 0.3 µmole/L) was significantly lower ($^{+}p<0.01$) than in men (3.6 ± 0.5 µmole/L). After methionine loading this sex difference persisted although the absolute rises in both were similar at all time intervals. Maximal levels were reached at 4 hr (7.0 ± 1.4 µmole/L in women and 11.1 ± 0.7 µmole/L in men, $^{+}p<0.01$).

*Effects of methionine loading in homocystinuria patients off pyridoxine treatment*

Fasting serum methionine levels in the patients ranged from 34 to 137 µmole/L, and the mean concentration, 76 ± 10 µmole/L, was significantly higher ($^{+}p<0.0001$) than in the group of control subjects (28 ± 2 µmole/L), although individual values showed an overlap (Fig. 1, Table 3). After methionine loading the mean peak level (1205 ± 86 µmole/L, range 849 to 1627) obtained at 1 hr, did not differ significantly from the mean in the 20 control subjects (942 ± 77 µmole/L, range 429 total 1487). The rate of decline calculated from the differences between the means at 1 hr (peak) and all time intervals thereafter did not differ between the groups of controls and patients.

The fasting homocystinemia in the patients ranged widely from as low as 1 to 74 µmole/L, with a mean value of 29. In every patient, however, methionine loading evoked a steady and significant rise of the homocystine concentration up to 73 ± 6 µmole/L (range 44-104) at 8 hr.

The mean fasting serum cystine level (29 ± 4 µmole/L) was significantly lower ($^{+}p<0.0001$) than in the 20 healthy control subjects (56 ± 2 µmole/L). Unlike the serum profile in the controls in all patients the serum cystine levels after methionine loading gradually fell ($^{++}p<0.001$) to a lowest mean value of 10 ± 1 µmole/L after 8 hr.

In the patients, the basal serum homocysteine-cysteine mixed disulphide levels were about 20 times higher than in the control subjects (38 ± 4 versus 2.2 ± 0.5 µmole/L). After methionine loading these levels remained virtually unchanged.

*Effects of methionine loading in homocystinuria patients on chronic high-dose pyridoxine treatment*

In the homocystinuria patients on chronic pyridoxine treatment the mean fasting serum methionine level was significantly lower than the mean level

*Table 3.* Individual concentrations of methionine, homocystine, cystine and homocysteine-cysteine mixed disulphide in serum before and 1, 2, 3, 4, 6, 8 hours after methionine loading (0.1 G/KG) in 8 patients with homocystinuria due to cystathionine synthase deficiency, without and during pyridoxine treatment

| Subjects | Without pyridoxine treatment: | | | | | | | During pyridoxine treatment: | | | | | | |
|---|---|---|---|---|---|---|---|---|---|---|---|---|---|---|
| hours | 0 | 1 | 2 | 3 | 4 | 6 | 8 | 0 | 1 | 2 | 3 | 4 | 6 | 8 |
| | methionine (µmol/l) | | | | | | | | | | | | | |
| Patient 1 | 71 | 1470 | 1067 | 917 | 879 | 725 | 750 | 30 | 1512 | 1212 | 880 | 742 | 677 | 599 |
| 2 | 80 | 1186 | 901 | 703 | 787 | 681 | 584 | 33 | 1417 | 1176 | 930 | 1050 | 752 | 698 |
| 3 | 69 | 1627 | 1105 | 1017 | 937 | 847 | 806 | 15 | 998 | 1127 | 938 | 817 | 718 | |
| 4 | 52 | 849 | 734 | 751 | 716 | 642 | 621 | 51 | 1030 | 1069 | 995 | 772 | 706 | 623 |
| 5 | 85 | 953 | 791 | 731 | 724 | 682 | 662 | 40 | 837 | 807 | 754 | 676 | 622 | 667 |
| 6 | 81 | 1307 | 1070 | 955 | 930 | 836 | 829 | 34 | 1224 | 1052 | 868 | 784 | 651 | 601 |
| 7 | 137 | 1159 | 1221 | 1116 | 1076 | 1164 | 893 | 37 | 1240 | 1102 | 1024 | 894 | 818 | 792 |
| 8 | 34 | 1091 | 766 | 707 | 606 | 562 | 823 | 24 | 1203 | 902 | 904 | 736 | 450 | 382 |
| Controls° | | | | | | | | | | | | | | |
| ♀ (n=10) | 22±2 | 854±101 | 765±70 | 728±63 | 647±61 | 549±58 | 453±54 | | | | | | | |
| ♂ (n=10) | 33±3** | 1030±107 | 802±51 | 694±43 | 626±44 | 472±40 | 392±40 | | | | | | | |
| ♀+♂ (n=20) | 28±2 | 942±77 | 784±43 | 711±39 | 636±38 | 510±37 | 423±35 | | | | | | | |
| | Homocystine (µmol/l) | | | | | | | | | | | | | |
| Patient 1 | 17 | 34 | 54 | 65 | 79 | 98 | 104 | 0 | 10 | 19 | 16 | 29 | 35 | 41 |
| 2 | 14 | 28 | 41 | 23 | 59 | 68 | 66 | 0 | 7 | 16 | 26 | 35 | 41 | 47 |
| 3 | 8 | 17 | 23 | 33 | 24 | 45 | 44 | 0 | 1 | 3 | 6 | 6 | 8 | |
| 4 | 40 | 45 | 54 | 61 | 53 | 72 | 78 | 1 | 8 | 20 | 31 | 34 | 53 | 66 |
| 5 | 21 | 35 | 38 | 51 | 41 | 71 | 73 | 3 | 4 | 11 | 21 | 23 | 35 | 50 |
| 6 | 74 | 80 | 68 | 69 | 71 | 67 | 69 | 0 | 6 | 15 | 21 | 24 | 37 | 36 |

| | | | | | | | | | | | | | | |
|---|---|---|---|---|---|---|---|---|---|---|---|---|---|---|
| | Homocystine (µmol/l) | | | | | | | | | | | | | |
| 7 | 60 | 61 | 73 | 79 | 83 | 96 | 91 | 0 | 6 | 17 | 24 | 29 | 43 | 46 |
| 8 | 1 | 12 | 19 | 27 | 33 | 45 | 58 | 0 | 1 | 3 | 5 | 4 | 9 | 8 |
| Controls° | | | | | | | | | | | | | | |
| ♀ (n=10) | ND | ND | 0.1± 0.1 | 0.1± 0.1 | 0.1± 0.1 | ND | 0.2± 0.2 | | | | | | | |
| ♂ (n=10) | ND | 0.2± 0.1 | 0.4± 0.1 | 0.5± 0.1 | 0.5± 0.2 | 0.4± 0.2 | 0.5± 0.2 | | | | | | | |
| ♀+♂ (n=20) | ND | ND | 0.3± 0.1 | 0.3± 0.1 | 0.3± 0.1 | 0.3± 0.1 | 0.3± 0.1 | | | | | | | |
| | Cystine (µmol/l) | | | | | | | | | | | | | |
| Patient 1 | 34 | 39 | 23 | 16 | 15 | 10 | 9 | 52 | 50 | 39 | 28 | 25 | 22 | 19 |
| 2 | 30 | 21 | 17 | 26 | 10 | 6 | 6 | 58 | 51 | 37 | 27 | 23 | 19 | 16 |
| 3 | 38 | 35 | 29 | 22 | 20 | 16 | 15 | 30 | 29 | 26 | 26 | 33 | 23 | |
| 4 | 20 | 13 | 11 | 11 | 9 | 6 | 6 | 58 | 48 | 35 | 27 | 23 | 16 | 15 |
| 5 | 33 | 27 | 20 | 15 | 14 | 11 | 9 | 62 | 57 | 43 | 33 | 29 | 23 | 13 |
| 6 | 10 | 10 | 6 | 9 | 10 | 7 | 7 | 43 | 44 | 32 | 23 | 21 | 16 | 13 |
| 7 | 22 | 20 | 17 | 15 | 13 | 11 | 12 | 64 | 57 | 46 | 38 | 31 | 23 | 21 |
| 8 | 48 | 45 | 36 | 28 | 25 | 17 | 13 | 51 | 60 | 51 | 51 | 48 | 43 | 36 |
| Controls° | | | | | | | | | | | | | | |
| ♀ (n=10) | 55±2 | 62± 2 | 61± 3 | 60± 2 | 60± 2 | 58± 2 | 53± 2 | | | | | | | |
| ♂ (n=10) | 58±2 | 62± 2 | 64± 3 | 64± 2 | 64± 2 | 60± 3 | 58± 3 | | | | | | | |
| ♀+♂ (n=20) | 56±2 | 64± 2 | 63± 2 | 62± 2 | 62± 2 | 59± 2 | 55± 2 | | | | | | | |
| | Homocysteine-cysteine mixed disulphide (µmol/l) | | | | | | | | | | | | | |
| Patient 1 | 48 | 60 | 65 | 65 | 67 | 74 | 69 | 8 | 27 | 34 | 35 | 44 | 44 | 42 |
| 2 | 34 | 37 | 40 | 34 | 43 | 41 | 39 | 9 | 23 | 35 | 40 | 42 | 40 | 43 |
| 3 | 28 | 34 | 39 | 40 | 38 | 41 | 41 | 4 | 11 | 20 | 21 | 21 | 24 | |
| 4 | 42 | 38 | 41 | 38 | 35 | 36 | 38 | 16 | 27 | 40 | 44 | 48 | 48 | 54 |

| | | | | | | | | | | | | | | |
|---|---|---|---|---|---|---|---|---|---|---|---|---|---|---|
| 5 | 40 | 51 | 43 | 44 | 39 | 46 | 46 | 11 | 21 | 35 | 42 | 44 | 51 | 53 |
| 6 | 41 | 42 | 35 | 39 | 40 | 25 | 24 | 6 | 18 | 25 | 29 | 30 | 30 | 28 |
| 7 | 55 | 50 | 52 | 52 | 50 | 55 | 47 | 6 | 23 | 38 | 42 | 42 | 43 | 45 |
| 8 | 12 | 31 | 35 | 39 | 42 | 39 | 43 | 3 | 9 | 16 | 22 | 23 | 27 | 25 |
| Controls° | | | | | | | | | | | | | | |
| ♀ (n=10) | 0.9±0.3 | 3.8± 0.9 | 6.0± 1.2 | | 7.0± 1.4 | | 7.0± 1.4 | | 6.7± 1.2 | | 6.9± 1.3 | | | |
| ♂ (n=10) | 3.6±0.5** | 6.7± 0.8* | 8.6± 1.1* | | 10.7± 0.7** | | 11.1± 0.7** | | 10.4± 1.4* | | 9.5± 1.3* | | | |
| ♀+♂ (n=20) | 2.2±0.4 | 5.3± 0.7 | 7.3± 0.9 | | 9.0± 0.9 | | 9.0± 0.9 | | 8.6± 1.0 | | 8.2± 1.0 | | | |

°Controls: mean concentrations ± S.E.M.
*$P < 0.05$ vs women, ** $P < 0.01$ vs women.

without treatment ($33 \pm 4$ versus $76 \pm 10$ µmole/L, $p<0.01$) and no longer differed from the mean fasting methionine level in the healthy control subjects ($28 \pm 2$ µmole/L) (Fig. 1, Table 3). After methionine loading the amino acid profile on chronic treatment essentially resembled the methionine curve without treatment but from 2 hr on the values at all time intervals were still higher than the corresponding levels in the healthy controls ($^{+}p<0.01$).

On chronic pyridoxine treatment the fasting serum homocystine levels became undetectable in 6 patients whereas the values in the remaining 2 decreased from 40 to 1 and from 21 to 3µmole/L. After methionine loading, however, marked homocystinemia reappeared at 1 hr and rose steadily thereafter with maximal values occurring at 8 hr (mean $42 \pm 6$µmole/L, range 8-66). Nevertheless, at all time intervals the homocystine levels were significantly lower than without pyridoxine treatment. The curves on and off treatment paralleled each other with an almost constant mean difference of about 30 µmole/L.

The mean fasting serum cystine level on chronic treatment was significantly higher ($p<0.05$) than the mean value without treatment ($52 \pm 4$ versus $29 \pm 4$ µmole/L) and similar to that in the healthy controls ($52 \pm 4$ versus $56 \pm 2$ µmole/L, $^{+}p>0.10$). Yet, on pyridoxine treatment, the serum cystine profile in the patients again showed – unlike in the control subjects – a clear-cut fall ($^{++}p<0.001$) to a minimal mean level of $19 \pm 3$ µmole/L (range 13-36) after 8 hr, at that time still significantly higher ($p<0.01$) than without treatment ($10 \pm 1$ µmole/L).

The mean serum level of homocysteine-cysteine mixed disulphide on chronic pyridoxine treatment was in the fasting state significantly lower than the value without treatment ($8 \pm 1$ versus $38 \pm 4$ µmole/L, $p<0.01$), although still significantly higher ($^{+}p<0.01$) than the mean fasting level in the healthy control subjects ($2.2 \pm 0.4$ µmole/L). After methionine loading a brisk rise ($^{++}p<0.001$) of the serum disulphide concentrations occurred so as to reach almost identically high levels after 8 hr ($41 \pm 4$ versus $43 \pm 4$ µmole/L) as without pyridoxine treatment.

*The levels of homocystine and homocysteine-cysteine mixed disulphide in serum during the day in two homocystinuria patients on chronic high-dose pyridoxine treatment and a regular diet*

During treatment with pyridoxine (750 mg/day) and on a normal diet in 2 patients the homocystine levels rose up to maximally 3.6 µmole/L and 13.3 µmole/L respectively, and the concentrations of homocysteine-cysteine mixed disulphide to 17.8 µmole/L and 31.9 µmole/L, respectively 9 and 5 hr after dinner time (Fig. 2). Even during the night there was some homocystinemia up to 6.0 µmole/L in one patient. The fasting levels of homocysteine-cysteine mixed disulphide in both patients were elevated (8.9 µmole/L and 18.0 µmole/L versus $0.9 \pm 0.3$ µmole/L in control women).

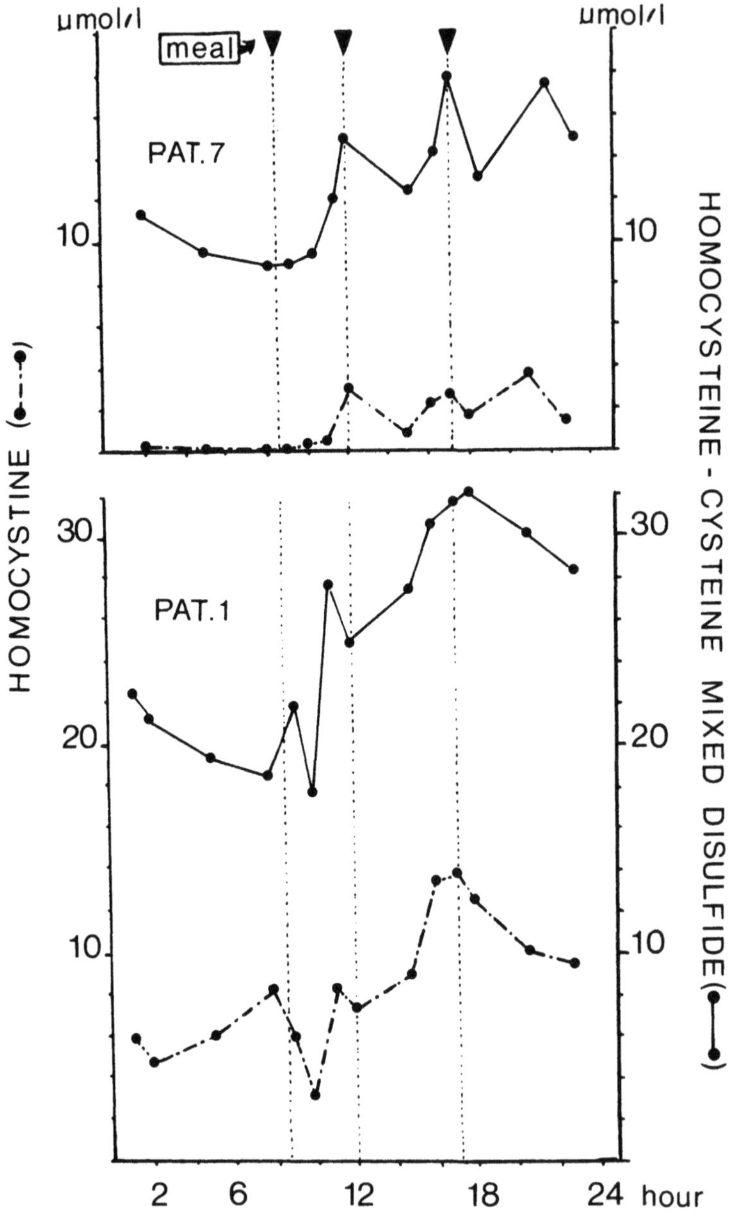


*Fig. 2* Diurnal serum concentrations of homocystine (●---●) and homocyteine-cysteine mixed disulphide (●—●) in 2 homocystinuria patients during pyridoxine treatment (750 mg/day) on a normal diet.

DISCUSSION

This study establishes that oral loading with methionine enables distinguishing of control subjects from adult patients with homocystinuria due to cystathionine synthase deficiency even on treatment with high-dose pyridoxine.

In adult female volunteers, the fasting levels of methionine and homocysteine-cysteine mixed disulphide were lower than in adult men, confirming earlier data.[20,21] After methionine loading the serum levels of this amino acid in men and women were similar. Nevertheless, the sex difference in the measured disulphide level persisted. The measured cystine levels – in women slightly, though not significantly, lower than in men – transiently increased in both groups after the methionine load. This ephemeral rise can be readily explained by prompt and active transsulphuration in both control men and women. Together, these data do not permit to draw firm conclusions about the cause of these sex differences.

In the group of untreated patients with homocystinuria due to proven cystathionine synthase deficiency the fasting methionine, homocystine and mixed disulphide levels were in the majority of the cases much higher and the measured cystine levels lower than in the control subjects. Interestingly, in one case (patient 8) with undoubtedly classical clinical signs of homocystinuria the fasting methionine and cystine levels were in the normal range with homocystinemia as low as 1.0 µmole/L. The raised fasting mixed disulphide level in this patient disclosed the presence of later proven cystathionine synthase deficiency.

After methionine loading with 0.7 mmole/kg body weight in this group of adult patients neither the peak levels of methionine nor the rate of decline during 8 hr thereafter differed significantly from our control group in contrast with literature data.[7,22,23] The measured homocystine levels steadily increased to more than twice the fasting concentrations and, concomitantly, the measured cystine levels decreased – in marked contrast to the transient increase in the control subjects – to approximately 33% of the preload values. The high mixed disulphide levels remained virtually unchanged. One could simply imagine that ongoing formation of the homocysteine-cysteine mixed disulphide in the presence of severely impaired transsulphuration of homocysteine to cysteine leads to the as yet unreported combination of increasing homocystine and decreasing cystine levels after loading with methionine. The virtually constant levels of the mixed disulphide after the methionine load suggest that the rate of its formation in this group of untreated patients with homocystinuria is near its maximum.

By chronic treatment with high-dose pyridoxine fasting levels of methionine and cystine normalized whereas homocystine was no longer detectable in the majority of the patients. The fasting disulphide concentrations remained elevated though at a much lower level than without treatment. Increased activity of cystathionine synthase by pyridoxal phosphate is the generally accepted mechanism to explain this supposedly therapeutic effect of the vitamin.[16]

After methionine loading the pattern of the serum levels of this amino acid did not differ significantly from that without treatment. Thus, the methionine load was similar off and on treatment. The patterns of the homocystine levels were roughly congruent during the 8 hr of observation, though at the average at an almost 30 µmole/L lower level. Concomitantly, the decrease of the cystine levels almost paralleled that without treatment whereas the disulphide concentrations – in contrast to the off treatment pattern – rose from much lower fasting levels to values similar to those off treatment at 8 hr after methionine loading. This combination of rising homocystine, decreasing cystine and increasing disulphide levels remembers of the findings by Rassin et al.[24] in normal subjects loaded with homocysteine. In their study however, the rise of homocystine and the increase of disulphide levels after loading with homocysteine was short lived and these changes had disappeared within 8 hr after loading. Thus, in our patients with proven cystathionine synthase deficiency, there is a much slower conversion of homocysteine to cysteine than in normal subjects despite high-dose pyridoxine treatment. Pyridoxine treatment, therefore, results merely in removal of the metabolic block between homocysteine and cysteine under circumstances of fasting. Comparing the data in this group of patients off and on treatment with pyridoxine, it is noteworthy that at levels of homocystine on treatment similar to those off treatment (confer 0 and 6 hr levels), the concurring cystine and disulphide levels are almost identical. This observation further indicates that on treatment with pyridoxine the rate of transsulphuration from homocysteine to cysteine after loading with methionine is essentially not differing from the off treatment rate. Patients with cystathionine synthase deficiency on treatment with high-dose pyridoxine are clearly still unable to metabolize methionine and homocysteine as normal subjects do.

In a number of clinical observations in our patients on high-dose pyridoxine treatment still marked homocystinemia and pathologically high mixed disulphide levels were found when they used a regular hospital diet (Fig. 2). Apparently the tiny increase of the cystathionine synthase activity during pyridoxine treatment as indicated by the results of our in vitro measurements in fibroblasts cultures might be responsible for the partial normalization of the fasting levels of methionine, homocystine, the mixed disulphide and the cystine levels but leaves the capacity to handle a major methionine load impaired. This conclusion is at variance with the supposition of Mudd et al. that a small percentage increase in the residual activity of the enzyme by pyridoxine treatment would be sufficient to metabolize dietary loads of methionine.[16] Restriction of major methionine loads from the diet and at least a more equal distribution of the methionine intake during the day, therefore, may be of additional value in defending these homozygous patients against the accumulation of homocysteine. This seems mandatory as it has been indicated in animal experiments[12,13] and very recently in cultured endothelial cells from normal humans[14] and from a heterozygous homocystinuric subject as well[15] that such accumulation is probably an important mediator of the widespread endothelial injuring that underlies a number of gross organ-damaging processes which play a key role in the spectrum of clinical abnormalities in this intriguing metabolic disorder.

REFERENCES


1. McKusick VA: Homocystinuria, in Heritable Disorders of Connective Tissue, ed. 4. St Louis, Mosby, 1972, pp. 224–281.
2. Barber GW, Spaeth GL: Pyridoxine therapy in homocystinuria. Lancet 1: 337, 1967.
3. Hooft C, Carton D, Samyn W: Pyridoxine treatment in homocystinuria. Lancet 1:1383, 1967.
4. Hollowell JG, Coryell ME, Hall WK, et al: Homocystinuria as affected by pyridoxine, folic acid and vitamin $B_{12}$. Proc Soc Exp Biol Med 129: 327–333, 1968.
5. Yoshida T, Tada K, Yokoyama Y, et al: Homocystinuria of vitamin $B_6$ dependent type. Tohoku J Exp Med 96: 235–242, 1968.
6. Barber GW, Spaeth GL: The successful treatment of homocystinuria with pyridoxine. J Ped 75: 463–478, 1969.
7. Gaull GE, Rassin DK, Sturman JA: Enzymatic and metabolic studies of homocystinuria: Effect of pyridoxine. Neuropaediatrie 1: 199–226, 1969.
8. Carson NA, Carré IJ: Treatment of homocystinuria with pyridoxine. Arch Dis Child 44: 387–392, 1969.
9. Mudd SH, Edwards WA, Loeb PM, et al: Homocystinuria due to cystathionine synthase deficiency: The effect of pyridoxine. J Clin Invest 49: 1762–1773, 1970.
10. Brenton DP, Cusworth DC: The response of patients with cystathionine synthase deficiency to pyridoxine, in Carson NA, Raine DN (eds): Inherited Disorders of Sulphur Metabolism. London, Churchill Livingstone, 1971, pp. 264–274.
11. Seashore MR, Durant JL, Rosenberg LE: Studies on the mechanism of pyridoxine-responsive homocystinuria. Pediatr Res 6: 187–196, 1972.
12. McCully KS, Wilson RB: Homocysteine theory of arteriosclerosis. Atherosclerosis 22: 215–227, 1975.
13. Harker LA, Ross R, Slichter SJ, et al: The role of endothelial cell injury and platelet response in its genesis. J Clin Invest 58: 731–741, 1976.
14. Wall RT, Harlan JM, Harker LA, et al: Homocysteine-induced endothelial cell injury in vitro: A model for the study of vascular injury. Thromb Res 18: 113–121, 1980.
15. de Groot PG, Willems Ch, Boers GH: Endothelial cell dysfunction in homocystinuria. Eur J Clin Invest 13: 405–410, 1983.
16. Mudd SH, Levy HL: Disorders of transsulphuration, in Stanbury JB (ed): Metabolic Basis of Inherited Disease. New York, McGraw Hill Book Company, 1978, pp. 458–503.
17. Drayer JI, Cleophas AJ, Trijbels JM, et al: Symptoms, diagnostic pitfalls, and treatment of homocystinuria in seven adult patients. Neth J Med 23: 89–94, 1980.
18. Fowler B, Kraus J, Packman S, et al: Evidence of three distinct classes of cystathionine betha-synthase mutants in cultured fibroblasts. J Clin Invest 61: 645–653, 1978.

19. Perry TL: Homocystinuria, in Nylan WL (ed): Heritable Disorders of Amino Acid Metabolism. New York, Wiley, 1974, pp. 395–428.
20. Armstrong MD, Stave U: A study of plasma free amino acid levels. II. Normal values for children and adults. Metabolism 22: 561–569, 1973.
21. Wilcken DE, Gupta VJ: Cysteine-homocysteine mixed disulphide: Differing plasma concentrations in normal men and women. Clin Sci 57: 211–215, 1979.
22. Brenton DP, Cusworth DC, Gaull GE: Homocystinuria: Metabolic studies in 3 patients. J Ped 67: 58–68, 1965.
23. Gaull GE, Sturman JA: Vitamin $B_6$ dependency in homocystinuria. Br Med J 3: 532–533, 1971.
24. Rassin DK, Longhi RC, Sternowsky HJ, et al: Homocysteine and cysteine loads in patients with homocystinuria due to cystathionine synthase deficiency: Effects of vitamin $B_6$. Clin Chim Acta 79: 197–210, 1977.

*Chapter 4*

# Homocystinuria versus Marfan's syndrome: the therapeutic relevance of the differential diagnosis

G.H.J. Boers, T.W. Polder, J.R.M. Cruysberg*, H.C. Schoonderwaldt**, J.J. Peetoom***, T.W.J. van Ruyven†, A.G.H. Smals and P.W.C. Kloppenborg

Summary

The clinical resemblance of homocystinuria to Marfan's syndrome often leads to a misdiagnosis of Marfan's syndrome in homocystinuria patients. The two diseases should be differentiated in any suspected patient in view of the risk of premature arteriosclerosis and thromboembolic complications in homocystinuria. Furthermore, in the case of homocystinuria the misdiagnosis deprives the patient of treatment with pyridoxine, folic acid and methionine restriction, which reportedly has a favourable biochemical and clinical effect in the majority of these cases. This relevance of the differential diagnosis is illustrated by the case histories of 2 homocystinuria patients. Quantitative serum amino acid analysis to exclude the possibility of homocystinuria should be performed in any patient suspected of Marfan's syndrome, especially when the mode of inheritance is not apparently autosomal dominant.

## INTRODUCTION

Homocystinuria, an inborn error of methionine metabolism with autosomal recessive inheritance, is caused in its most frequent form by deficiency of cystathionine β-synthase activity. This enzyme transsulphurates homocysteine – the metabolite of the essential amino acid methionine – to cystathionine which is converted to cysteine (fig. 1). Homocystinuria is also a feature

Department of Internal Medicine, Division of Endocrinology; *Department of Ophthalmology; **Department of Neurology, St Radboud Hospital, University of Nijmegen; ***Department of Internal Medicine, Municipal Hospital, Arnhem; †Institute for Mentally Retarded Patients 'De Hardenberg', Ede.

*Neth J Med 1984; 27: 206-12.*

of two other, extremely rare inherited metabolic defects: 5-methyltetrahydrofolate-homocysteine methyltransferase deficiency due to defective production of co-factor methylcobalamin or 5,10-methylenetetrahydrofolate reductase deficiency. These defects lead to a decreased remethylation rate of homocysteine to methionine. Furthermore, homocystinuria has been observed in the vitamin $B_{12}$ malabsorption syndrome (Imerslund-Gräsbeck syndrome) and in vitamin $B_6$ depletion[1].

In classical homocystinuria, due to cystathionine β-synthase deficiency,

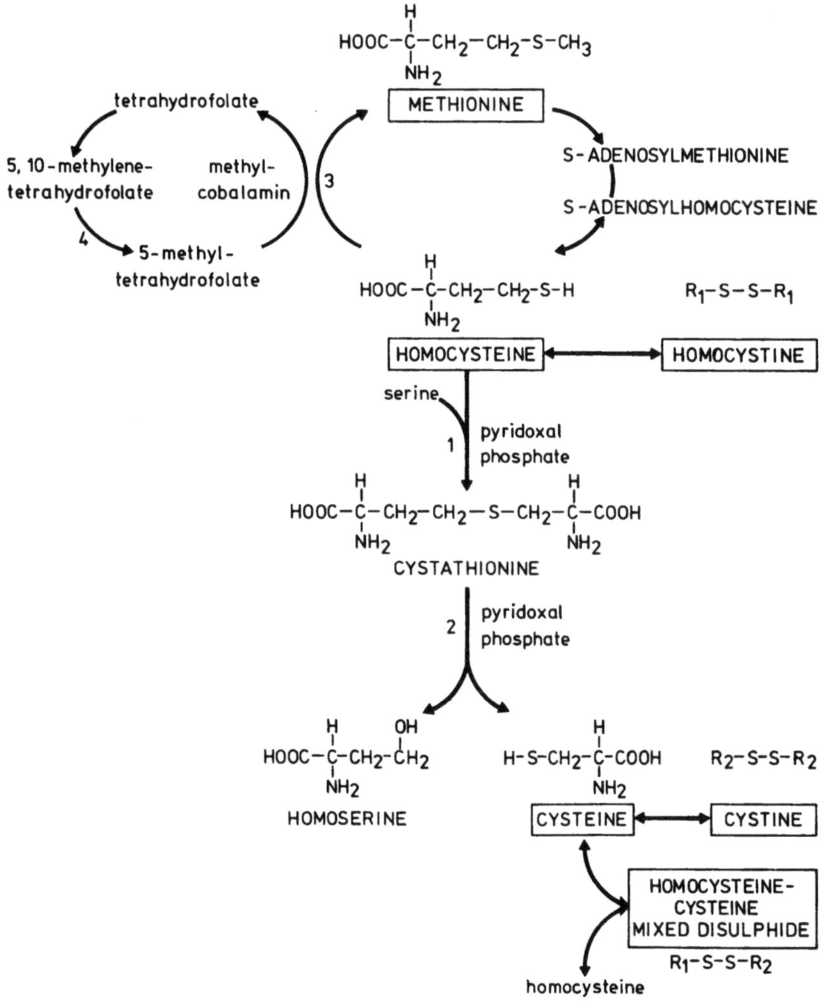


*Fig. 1* The metabolism of methionine. 1 = cystathionine β-synthase; 2 = γ-cystathionase; 3 = 5-methyltetrahydrofolatehomocysteine methyltransferase; 4 = 5,10-methylene tetrahydrofolate reductase.

methionine and homocysteine accumulate in tissues and blood and are excreted in large amounts in the urine, whereas cysteine levels are decreased. Homocystinemia is supposedly the factor causing the symptoms of the syndrome. The clinical picture is very heterogeneous, with as most frequent symptoms: ectopia lentis, premature arteriosclerosis and serious thromboembolic complications, osteoporosis, scoliosis, dolichostenomelia (excessively long and thin extremities), arachnodactyly, pectus carinatum or excavatum, mental retardation, and neurologic and psychiatric disorders. The estimated incidence is 1:24,000-200,000, dependent on geographical location and screening method. The diagnosis is established by serum amino acid analysis and by determination of cystathionine β-synthase activity in a fibroblast culture obtained by skin biopsy. In homozygotes the activity of the enzyme is absent or residual, i.e. about ½% of mean normal activity[2]. Treatment with pyridoxine gives a favourable biochemical and clinical response in about 50 per cent of the patients reported so far[2,3]. Pyridoxal phosphate, co-factor of cystathionine β-synthase, is supposed to enhance the residual enzyme activity in these patients when administered in large doses.

Marfan's syndrome, a disorder of connective tissue with autosomal dominant inheritance, has been assumed to be an inborn error of protein metabolism, particularly in collagen or elastin or in both[4], but direct evidence is not yet available. In its classical form the syndrome is characterized by ectopia lentis, aneurysm of the aorta with high risk of dissection and rupture, prolapse of the mitral valve, excessive height, scoliosis, dolichostenomelia, arachnodactyly, pectus excavatum or carinatum, and joint laxity. The incidence is 1:17,000-25,000 and about 15 per cent of the cases are due to new mutations[5]. Since chemical abnormalities in Marfan's syndrome are not known so far, the disease can only be identified clinically.

In view of the clinical resemblance of both syndromes, homocystinuria patients are often misdiagnosed as suffering from Marfan's syndrome[6], and treatment with a possible favourable effect is then withheld. This should be avoided by serum amino acid analysis of any person suspected of having the Marfan's syndrome, especially when an autosomal dominant mode of inheritance is not apparent. In the 2 patients described here, the relevance of differentiating these disorders is illustrated.

## METHODS

Measurements of methionine, homocystine, homocysteine-cysteine mixed disulphide, and cystine concentrations in fasting serum were performed by quantitative amino acid analysis (Biotronik LC 2000). Serum was de-proteinized within 10 minutes after venipuncture. The levels of homocystine and homocysteine-cysteine mixed disulphide reflect the concentration of homocysteine in vivo, due to rapid oxidation after venipuncture of the latter to homocystine (homocysteine-homocysteine) and homocysteine-cysteine mixed disulphide[7].

Determinations of cystathionine β-synthase activity in liver obtained at autopsy in patient 1 and in a fibroblast culture obtained by skin biopsy in patient 2 were performed by Dr B. Fowler, Willink Biochemical Genetics Unit, Royal Children's Hospital, Manchester, using a method previously published[2,8].

## CASE HISTORIES

*Patient 1*, a female teacher born in 1960 from consanguineous parents, had an uneventful history until 1977 when subluxation of both eye lenses was discovered. In view of her height (1.85 m), dolichostenomelia, thoracic kyphosis and pectus excavatum, Marfan's syndrome was diagnosed. No other members of her family had signs of this syndrome. From 1978 on she complained of migraine-type headaches. In February 1980 she started taking oral contraceptives (Ministat, 0.75 mg lynestrenol and 0.0375 mg ethinyloestradiol). In July 1980, total luxation of the right eye lens into the anterior chamber causing pupillary block glaucoma was diagnosed. The lens of the left eye showed downward subluxation, as before (figs. 2, 3). The totally dislocated lens was replaced behind the iris by pressure on the cornea. Four days later she was admitted to our hospital for further evaluation. Now the right eye lens was situated in the corpus vitreum and there was minor right-sided hemiparesis with stiffness of the neck, which soon improved. But 5 days after admission she acutely entered coma with left-sided hemiplegia. The results of carotid angiography and a CT-scan of the brain disclosed temporal infarction with brain edema. In addition, angiography showed extreme narrowing of the common and cervical internal and intracranial carotid artery. On the strength of this finding in combination with the downward dislocation of the eye lenses, the diagnosis homocystinuria rather than Marfan's syndrome was suspected. Indeed, serum amino acid analysis gave biochemical evidence of homocystinuria due to cystathionine β-synthase deficiency (table 1). Pyridoxine treatment (4 × 1 g daily parenterally) was started, but the course was dramatic; one day after the onset of coma an epileptic state developed and artificial mechanical respiration was required; 2 days later she died. Postmortem examinations showed haemorrhagic infarction in the right superior temporal gyrus of the brain with edema and multiple thrombi in the superior sagittal sinus and in several leptomeningeal and pial veins. There were also old and recent infarctions of the renal parenchyma and recent infarction in the cardiac septum and left ventricular myocardium. These postmortem findings and the neurologic course have been previously published in detail[10]. In liver obtained by autopsy only residual activity of cystathionine β-synthase was measured (table 1), proving the diagnosis ho-

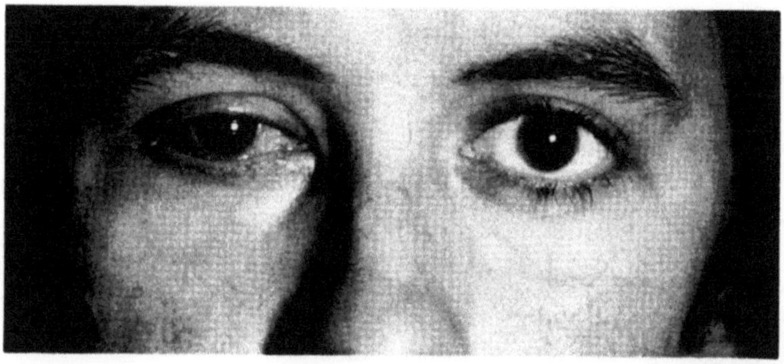

*Fig. 2.* The right eye lens is totally dislocated into the anterior chamber, causing acute pupillary block glaucoma; the left eye lens is subluxated downward in patient 1.

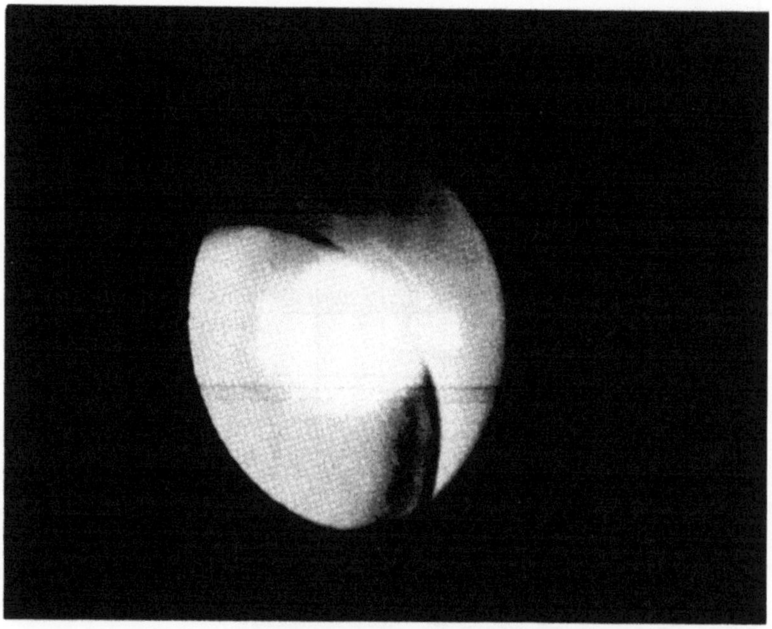

*Fig. 3.* Detail of fig. 2: the left eye lens is subluxated downward and the zonulae fibres are disrupted and tortuously retracted to the lens capsule in patient 1.

mocystinuria due to cystathionine β-synthase deficiency. The presence of some residual enzyme activity indicated that treatment with vitamin $B_6$ during her life could have prevented these fatal events.

*Patient 2*, a man born in 1967 from non-consanguineous parents, was diagnosed as suffering from ectopia lentis at the age of 1 year and as

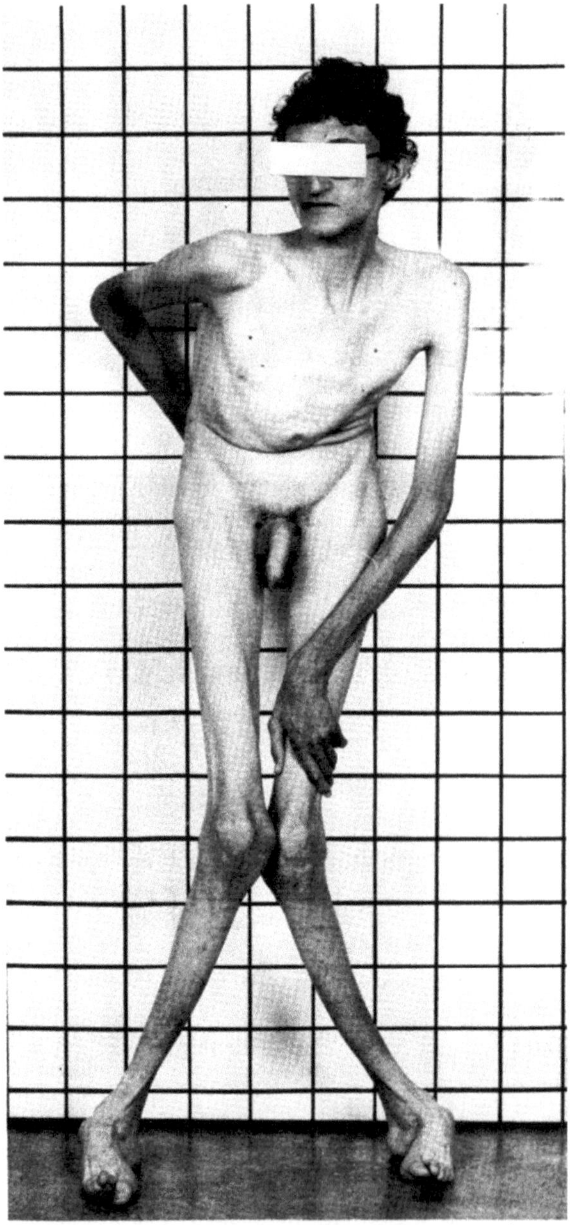

*Fig. 4.* Marfanoid habitus (dolichostenomelia, genua valga, pes equinovarus) in patient 2.

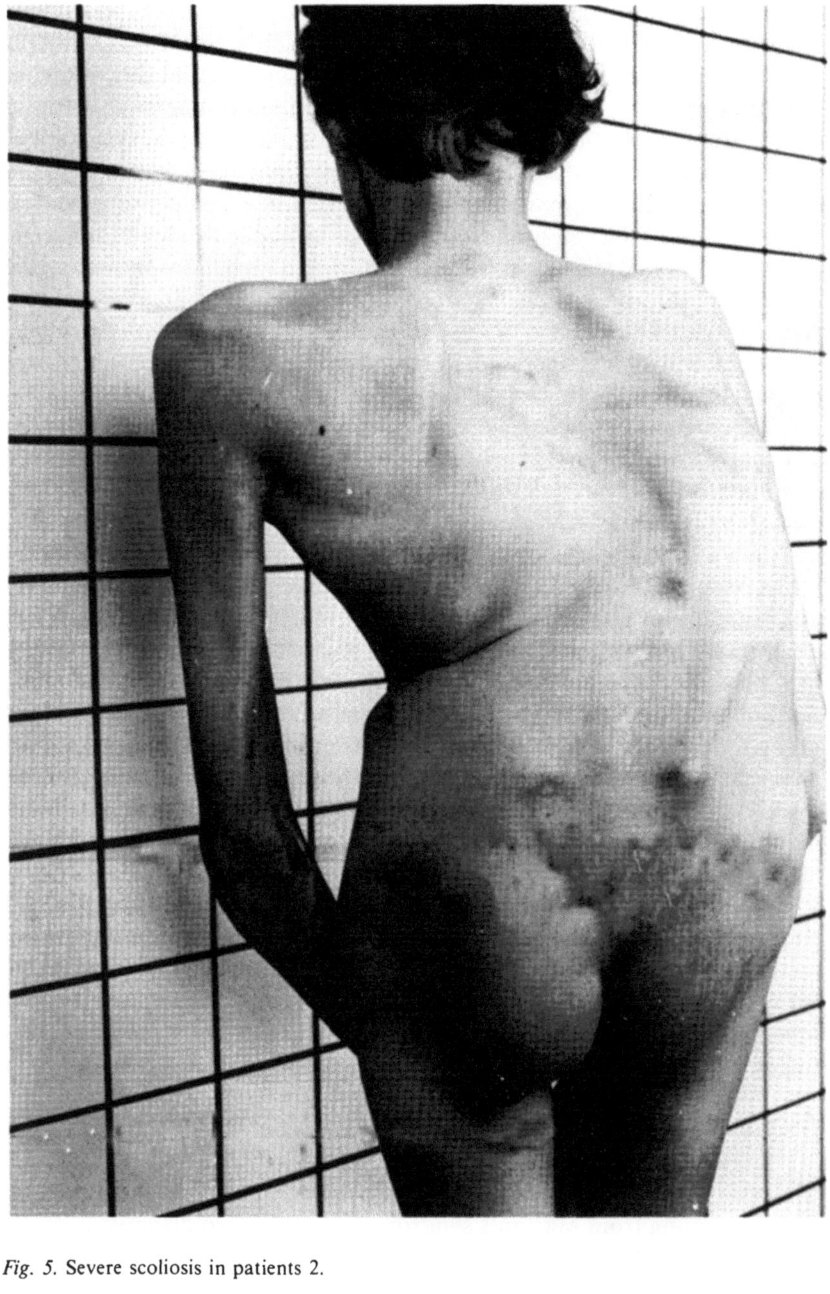

*Fig. 5.* Severe scoliosis in patients 2.

oligophrenic at the age of 2 years in view of motor and mental retardation. Evaluation at the age of 6 disclosed scoliosis, dolichostenomelia, arachnodactyly, genua valga, pes equinovarus, heart murmur, and downward subluxation of both eye lenses. Marfan's syndrome was diagnosed although oligophrenia (IQ 41, Stutman Intelligence test) is not part of the syndrome. In no other member of his family the syndrome was suspected. Because of aggravation of the scoliosis complicated by a significant reduction of the vital capacity of the lungs, he was referred to an orthopaedic clinic for surgical correction in 1982. His height was then 1.87 m and his habitus showed typical marfanoid features (figs. 4, 5). Preoperative echocardiographic examination disclosed holosystolic prolapse of the mitral valve. Ophthalmological evaluation now showed lenses totally dislocated into the vitreous with disrupted zonulae fibres. In view of this finding in combination with his oligophrenia, the diagnosis of Marfan's syndrome was rejected and homocystinuria was suspected. Amino acid analysis indeed confirmed the diagnosis homocystinuria due to cystathionine β-synthase deficiency (table 1). Determination of the enzyme activity in a fibroblast culture obtained by skin biopsy did not show any activity at all before and after addition of pyridoxine phosphate to the cell extract (table 1). Homocystinuria due to cystathionine β-synthase deficiency, probably not responding to pyridoxine treatment in vivo, was established. Surgery was postponed in view of the reported great risk of thromboembolic complications of anaesthesia, operation and immobilization in homocystinuria patients. Treatment with pyridoxine (750 mg daily orally for 6 weeks) as expected had no favourable effect on the amino acid abnormality, but additional treatment with folic acid (15 mg daily orally for 6 weeks) substantially reduced the homocystinemia (table 1). Enhanced re-methylation of homocysteine to methionine (fig. 1) must have accounted for this effect in view of the increase of methionine and the unchanged low cystine level during this treatment. Then, in addition to pyridoxine and folic acid treatment, a methionine-restricted diet during 6 weeks, containing 600 mg methionine in 80 g proteins, reduced the homocystinemia to levels acceptable for surgical correction of the scoliosis (table 1). The postoperative course and immobilization in a plaster brace during 8 months thereafter were uneventful, while oral anticoagulant therapy was given during this period.

## DISCUSSION

This study reports on two patients with homocystinuria, misdiagnosed earlier as suffering from Marfan's syndrome, a well-known diagnostic error in these patients[6]. The relevance of differentiating these two diseases has been pointed out before[11], and in patient 1 the dramatic clinical outcome caused by thromboembolic complications could probably have been prevented by appropriate diagnosing and treatment. Assessment of cystathionine β-synthase

*Table 1.* Fasting serum amino acid concentrations before and during treatment in µmol/l and cystathionineβ-synthase activity in nmol/h/mg protein in patients 1 and 2

| | | Patient 1 | Patient 2 | Normal range[9] (n=20) |
|---|---|---|---|---|
| Methionine | | | | |
| untreated | | 48 | 549 | 16 - 47 |
| treated with* | pyridoxine | | 655 | |
| | pyridoxine and folic acid | | 974 | |
| | pyridoxine and folic acid and methionine restriction | | 118 | |
| Homocystine | | | | |
| untreated | | 109 | 87 | ND** |
| treated with | pyridoxine | | 71 | |
| | pyridoxine and folic acid | | 43 | |
| | pyridoxine and folic acid and methionine restriction | | 25 | |
| Homocysteine-cysteine | | | | |
| untreated | | 56 | 30 | 0.2 - 5.6 |
| treated with | pyridoxine | | 32 | |
| | pyridoxine and folic acid | | 31 | |
| | pyridoxine and folic acid and methionine restriction | | 34 | |
| Cystine | | | | |
| untreated | | 6 | 4 | 47 - 67 |
| treated with | pyridoxine | | 6 | |
| | pyridoxine and folic acid | | 8 | |
| | pyridoxine and folic acid and methionine restriction | | 31 | |
| Cystathionine β-synthase activity | | | | |
| before pyridoxal phosphate addition to cell extract | | 0.54*** | ND† | 3.6 - 25.7† |
| after pyridoxal phosphate addition to cell extract | | 0.71*** | ND† | 4.9 - 32.1† |

* each treatment for a period of 6 weeks

** not detectable

*** in liver tissue obtained at postmortem examination, level in control liver 45.7 nmol/h/mg protein before and 66.8 nmol/h/mg protein in after addition of pyridoxal phosphate to the cell extract

† in fibroblast cultures by skin biopsy.

activity in liver obtained at a postmortem examination on this patient revealed a small residual activity of the enzyme, suggesting probable responsiveness to pyridoxine treatment in vivo. Furthermore, homocystinuric women should avoid oral contraceptives since estrogenic steroids may induce pyridoxine and folate deficiency[12,13]. Such pyridoxine deficiency can be catastrophic in a pyridoxine responsive homocystinuria patient due to a further decrease in residual cystathionine β-synthase activity. Additional folate defi-

*Table 2.* Comparison of homocystinuria and Marfan's syndrome

| | Homocystinuria | Marfan's syndrome |
|---|---|---|
| Inheritance | autosomal recessive | autosomal dominant |
| Ectopia lentis | usually downward with disrupted zonulae fibres, not congenital | usually upward with stretched non-disrupted zonulae fibres, usually congenital |
| Vascular disease | premature arteriosclerosis and high incidence of thromboembolic events | aneurysm, dissection, and rupture of the aorta, prolapse of the mitral valve |
| Skeletal abnormalities | osteoporosis with biconcavity and flattening of the vertebrae, dolichostenomelia, arachnodactyly, scoliosis, pectus excavatum or carinatum | large vertebrae, dolichostenomelia, arachnodactyly, scoliosis, pectus excavatum or carinatum, joint laxity |
| Neurologic abnormalities | mental retardation, convulsions, psychiatric illnesses | absent |

ciency in such a patient aggravates homocystinemia due to decreased re-methylation of homocysteine to methionine for lack of 5-methyltetrahydrofolate, which acts as methyl donor in this reaction (fig. 1). In this patient, the use of oral contraceptives in the 5 months preceding her death may indeed have contributed to the fatal course. Gröbe reported an analogous case history of a 19-year-old homocystinuric woman who developed major symptoms of the disease after starting oral contraceptives[14].

In patient 2 a diagnosis of Marfan's syndrome was erroneously made despite intensive examinations at ages 2 and 6 years. Institution of homocysteine-lowering treatment with folic acid and dietary methionine restriction from that age on might have prevented the development of his serious symptoms. Homocystinuria was diagnosed at the age of 15 years, when major surgery and protracted immobilization was planned. Anaesthesia, surgery, and immobilization are discouraged in homocystinuric patients in view of an excessive risk of fatal thromboembolic complications[15,16]. If absolutely indicated in view of pulmonary complications of the severe scoliosis as in our patient, anaesthesia and surgery should be performed under special precautions to prevent hypotension, dehydration, and stasis of blood; and anticoagulant treatment[17,18] is required. Attempts have to be made in advance to ensure maximal reduction of the homocystine concentration, because this has been reported to be linearly related to homocysteine-induced endothelial lesions and platelet adhesion to exposed subendothelium[19,20]. These measures may have contributed to the prevention of thromboembolic events during the major surgical procedure and the 8 months of immobilization thereafter.

Differentiating Marfan's syndrome from homocystinuria solely on the basis of the clinical symptoms may be very hazardous. Admittedly, there are several differences in the clinical features of both syndromes, but none of these is absolute (table 2). Thus, in patient 1 the assessment of normal intelligence in combination with marfanoid habitus and ectopia lentis – although downward – erroneously led to a diagnosis of Marfan's syndrome; and in patient 2 the prolapse of the mitral valve in a marfanoid person with ectopia lentis – although downward – was also thought to be a strong argument for a diagnosis of Marfan's syndrome despite his oligophrenia, which was considered to be merely coincidental.

The outcome in these 2 patients illustrates the relevance of differentiating homocystinuria from Marfan's syndrome. The urinary cyanide-nitroprusside (Brand) test, being false-negative in some homocystinuric patients[3] and false-positive in many normal subjects, is not a valid screening method for homocystinuria. Quantitative serum amino acid analysis is therefore indicated in any patient suspected of homocystinuria, especially in patients considered to have Marfan's syndrome without an apparent autosomal dominant mode of inheritance.

REFERENCES


1. Mudd SH, Levy HL. Disorders of transsulphuration. In: Stanbury JB, Wyngaarden JB, Fredrickson DS, eds. The metabolic bases of inherited disease. New York: McGraw-Hill, 1983.
2. Boers GHJ, Smals AGH, Drayer JIM, Trijbels JMF, Leermakers AI, Kloppenborg PWC. Pyridoxine treatment does not prevent homocystinemia after methionine loading in adult homocystinuria patients. Metabolism 1983; 32: 390–7.
3. Drayer JIM, Cleophas AJM, Trijbels JMF, Smals AGH, Kloppenborg PWC. Symptoms, diagnostic pitfalls and treatment of homocystinuria in seven adult patients. Neth J Med 1980; 23: 89–94.
4. Abraham PA, Perejda AJ, Carnes WH, Uitto J. Marfan syndrome, demonstration of abnormal elastin in aorta. J Clin Invest 1982; 70: 1245–52.
5. Pyeritz RE, McKusick VA. The Marfan syndrome: diagnosis and management. N Engl J Med 1979; 300: 772–7.
6. McKusick VA. Homocystinuria. In: McKusick VA, ed. Heritable disorders of connective tissue.St Louis: Mosby, 1972: 224–81.
7. Perry TL, Homocystinuria. In: Nyan WL, ed. Heritable disorders of amino acid metabolism. New York: Wiley, 1974: 395–428.
8. Fowler B, Kraus J, Packman S, Rosenberg LE. Homocystinuria. Evidence of three distinct classes of cystathionine beta-synthase mutants in cultured fibroblasts. J Clin Invest 1978; 61: 645–53.
9. Boers GHJ, Smals AGH, Trijbels JMF, Leermakers AI, Kloppenborg PWC. Unique efficiency of methionine metabolism in premenopausal women may protect against vascular disease in the reproductive years. J Clin Invest 1983; 72: 1971–6.
10. Schoonderwaldt HC, Boers GHJ, Cruysberg JRM, Schulte BPM, Slooff JL, Thijssen HOM. Neurologic manifestations of homocystinuria. Clin Neurol Neurosurg 1981; 83: 153–62.
11. Brenton DP, Dow CJ, James JIP, Ray RL, Wynne-Davies R. Homocystinuria and Marfan's syndrome. A comparison. J Bone Joint Surg 1972; 54 B: 277–98.
12. McCully KS. Homocystine, atherosclerosis and thrombosis: implications for oral contraceptive users. Am J Clin Nutr 1975; 28: 542–9.
13. Rose DP, Strong R, Adams PW, Harding PE. Experimental vitamin $B_6$ deficiency and the effect of oestrogen-containing oral contraceptives on tryptophan metabolism and vitamin $B_6$ requirements. Clin Sci 1972; 42: 465–77.
14. Gröbe H. Homocystinuria and oral contraceptives. Lancet 1978; 1: 158–9.
15. Brown RB, Watson PD, Taussig LM. Congenital metabolic diseases of pediatric patients: anesthesiological implications. Anesthesiology 1975; 43: 197–209.
16. Regenbogen L, Ilie S, Elian I. Homocystinuria – a surgical and anesthetic risk. Metab Pediatr Ophthalmol 1980; 4: 209–12.
17. Crooke JW, Towers JF, Taylor WH. Management of patients with homocystinuria requiring surgery under general anaesthesia. Br J Anaesth 1971; 43: 96–9.
18. Parris WCV, Quimby CW. Anesthetic considerations for the patient with homocystinuria. Anesth Analg (Cleve) 1982; 61: 708–10.
19. Harker LA, Slichter SJ, Scott CR, Ross R. Homocystinemia. Vascular injury and arterial thrombosis. N Engl J Med 1974; 291: 537–43.
20. Harker LA, Ross R, Slichter SJ, Scott CR. Homocystine-induced arteriosclerosis, the role of endothelial cell injury and platelet response in its genesis. J Clin Invest 1976; 58: 731–41.

# Part II
# Heterozygous homocystinuria

*Chapter 5*

# Heterozygosity for homocystinuria

## 5.1 FREQUENCY OF HETEROZYGOSITY

The rate of detection of homozygous homocystinuria due to cystathionine synthase deficiency in newborn screening programs (Chapter 1.9) indicates that the frequency of heterozygotes for this condition with autosomal recessive inheritance might be as high as 1 in 70 to 1 in 200 in various populations[1]. Data on frequencies from screening of groups of normal subjects for heterozygosity for homocystinuria, however, are not available as yet. The lack of a valid test to differentiate heterozygotes from normal subjects is held to be an obstacle for such screening. In Chapter 7 it will be discussed that the relatively simple procedure of a standardized oral methionine loading test[2] is a valid method of detection of heterozygotes, under the proviso of differentiating the tested subjects according to sex and menopausal state. This condition is based upon the finding of small, but striking and statistically significant differences in methionine handling between pre- and postmenopausal women and men, as will be shown in detail in Chapter 6.
With this test we investigated 40 healthy control subjects and detected pathological methionine-handling in one adult woman. Heterozygosity was confirmed by determination of cystathionine synthase activity in her cultured fibroblasts which appeared to be as low as 17% of the mean enzyme activity in the fibroblasts of control subjects as measured by Dr. Fowler in Manchester. Nevertheless, screening of a group of normal control subjects large enough to estimate the frequency of heterozygosity for homocystinuria in the population in the Netherlands, has not been conducted by us until now.

## 5.2 MORBIDITY IN HETEROZYGOTES

Obligate carriers for homocystinuria, parents or children of homozygous

homocystinurics, reportedly show cystathionine synthase activities in their cultured skin fibroblasts to levels of 21 to 45% of the mean enzyme activity in normal control subjects[3]. By our group very recently a mean activity of 19% in 20 obligate heterozygotes has been reported (Chapter 7). Such carriers also demonstrate a higher degree of homocysteinemia after a methionine load than normal control subjects do (Chapter 7)[2-4]. In animal experiments and in in vitro studies on cultured endothelium cells the homocysteine-induced endothelium injury – leading to striking premature arteriosclerosis and thromboembolic events in homozygous patients (Chapter 1.6.3 and 2.6.3) – proved to be linearly related to the homocysteine concentrations in the blood of these animals or in the medium of the cell cultures (Chapter 1.7.2).
From these findings it was intriguing to explore the possibility whether mild homocysteinemia, as present in obligate heterozygotes for homocystinuria, may predispose to the development of complications typical for the homocystinuria syndrome, in particular the vascular symptoms. Wilcken et al[5] in 1976 reported that in 7 out of a group of 25 patients with angiographically proven coronary artery disease under the age of 50, the levels of homocysteine-cysteine mixed disulphide in plasma 4 hours after an oral methionine load were comparable with those measured by other investigators[2] in obligate heterozygotes for homocystinuria after similar loading. This finding suggested that even mild homocysteinemia might play a major role in the pathogenesis of vascular disease at early age. In 1983 however, this Australian group reported that by means of methionine loading they could not confirm a predominance of heterozygotes for homocystinuria among patients with premature coronary artery disease[4]. Only 2 of the 20 patients showed levels of total homocysteine after the methionine load exceeding the range in controls and comparable to those levels measured in 17 obligate heterozygotes. It has to be admitted however, that these authors did not differentiate their tested patients, obligate heterozygotes, and normal control subjects according to sex and menopausal state. By not doing so, they were unable to differentiate the obligate heterozygotes from the normal controls as 10 of 17 carriers showed after load levels of total homocysteine within the range of the control subjects, as will be discussed in detail in Chapter 6.
In an attempt to directly establish the incidence of heart attaque and stroke in a group of heterozygotes for homocystinuria, Mudd et al. analysed questionnaires from a large number of carriers[1]. Based upon these rather anecdotic data they estimated that no more than 5% of homocystinuria heterozygotes might have a fatal or non-fatal heart attaque by the age of 50. These data made a relative risk of heart attaque in these carriers higher than threefold improbable. Furthermore, no marked difference in the probability to develop strokes was detected by comparing these heterozygotes with their control group. Swift et al[6], however, stated on the strength of the same data that there are indications that substantial excess risk of cardiovascular disease or stroke might be associated with heterozygosity for the homocystinuria gene.

Challenged by these contradicting data from literature, we started in 1980 to explore the presence or absence of disturbed methionine metabolism due to heterozygosity for homocystinuria in groups of patients suffering from premature arteriosclerosis. Preliminary data from these studies have been reported before[7,8]. In Chapter 8 more definitive data will be presented. A frequency of heterozygosity as high as about 30% could be established in groups of patients suffering from premature occlusive peripheral or cerebrovascular arterial disease, whereas, remarkably, in a group of patients with myocardial infarction under the age of 50 years a frequency higher than in the general population so far has not been confined. The validity of the methods used to detect heterozygotes for homocystinuria has been shown in a previous report by us (Chapter 7)[9]. From these data we concluded that heterozygosity for homocystinuria leading to mild homocysteinemia indeed has to be considered as a risk factor of premature arteriosclerosis especially in peripheral and cerebrovascular vessels.

REFERENCES


1. Mudd SH, Havlik R, Levy HL, McKusick VA, Feinleib M. A study of cardiovascular risk in heterozygotes for homocystinuria. Am J Hum Genet 1981; 33: 883–93.
2. Sardharwalla IB, Fowler B, Robins AJ, Komrower GM. Detection of heterozygotes for homocystinuria. Arch Dis Child 1974; 49: 553–9.
3. Mudd SH, Levy HL. Disorders of transsulphuration. In: Stanbury JB, Wyngaarden JB, Fredrickson DS, eds. The metabolic basis of inherited disease. New York: McGraw-Hill Book Co, ed. 5, 1983; 522–59.
4. Wilcken DEL, Reddy SG, Gupta VJ. Homocysteinemia, ischemic heart disease, and the carrier state for homocystinuria. Metabolism 1983; 32: 363–9.
5. Wilcken DEL, Wilcken B. The pathogenesis of coronary artery disease. A possible role for methionine metabolism. J Clin Invest 1976; 57: 211–5.
6. Swift M, Morrell D. Cardiovascular risk in homocystinuria family members. Am J Hum Genet 1982; 34: 1016–8.
7. Boers GHJ, Schoonderwaldt HC, Schulte BPM, Trijbels JMF, Smals AGH, Kloppenborg PWC. Heterozygosity for homocystinuria: a risk factor of occlusive cerebrovascular disease? Clin Genet 1983; 24: 300–1.
8. Boers G, Trijbels J, Smals A, Kloppenborg P. Homocystinuria in heterozygous form as a risk factor of premature arteriosclerosis. Eur J Clin Invest 1984; 14: no 2-II: 9.
9. Boers GHJ, Fowler B, Smals AGH, Trijbels JMF, LeermakersAI, Kleijer WJ, Kloppenborg PWC. Improved identification of heterozygotes for homocystinuria due to cystathionine synthase deficiency by the combination of methionine loading and enzyme determination in cultured fibroblasts. Hum Genet 1985; 69: 164–9.

*Chapter 6*

# Unique efficiency of methionine metabolism in premenopausal women may protect against vascular disease in the reproductive years

Godfried H Boers, Anthony G. Smals, Frans J. Trijbels, Annelies I. Leermakers and Peter W. Kloppenborg

*Departments of Medicine, Division of Endocrinology and Pediatrics, University of Nijmegen, The Netherlands*

ABSTRACT

Premenopausal women develop occlusive artery disease less frequently than postmenopausal women. In coronary heart disease, higher blood levels of homocysteine-cysteine mixed disulphide have been reported. Therefore, in healthy subjects, we studied the role of menopausal status in the transsulphuration of methionine in 10 premenopausal and 10 postmenopausal women. To exclude the role of aging, we compared these results with those in 10 younger and 10 older men of comparable age groups. An oral methionine load (0.1 g/kg of body weight) was administered after overnight fasting. Before and during 8 h, thereafter, serum levels of methionine, homocystine, and homocysteine-cysteine mixed disulphide were measured. In the fasting state, serum methionine levels were similar in the premenopausal women and both groups of men. Postmenopausal women had significantly lower fasting levels. Peak levels and clearances of methionine after loading did not differ between the groups. In the fasting state, homocystine was never detectable; yet, after methionine loading, slight homocystinemia was present in 12 out of 20 men, and was more pronounced in all postmenopausal women. However, homocystinemia did not occur in any of the premenopausal women after loading. Fasting serum homocysteine-cysteine mixed disulphide levels did not differ between both groups of men and postmenopausal women. In premenopausal women, both fasting and postloading disulphide levels were significantly lower than in any other group. We conclude that premenopausal women have a unique efficiency of methionine handling, and thereby are preserved against the accumulation of homocysteine after methionine loading. We speculate that this phenomenon might account for the lower incidence of vascular disease in women in the reproductive life cycle.

*J Clin Invest 1983; 72: 1971-6.*

## INTRODUCTION

In man, the essential sulphur amino acid methionine is metabolized in the transsulphuration pathway[1]. Successively, it is converted to S-adenosyl-methionine, S-adenosyl-homocysteine, and homocysteine. Homocysteine lies at a branch point from which sulphur metabolism can be controlled: either it can be remethylated to methionine or converted to cysteine via cystathionine.

In homocystinuria, an inborn error of methionine metabolism that results from a decreased activity of the enzyme cystathionine β-synthase (EC 4.2.1.22) which normally converts homocysteine into cystathionine, homocysteine accumulates in the blood and tissues and is excreted in the urine. The presence of homocystine and of increased levels of homocysteine-cysteine mixed disulphide in these patients reflects accumulation of homocysteine since the latter is readily oxidized and measured as homocystine (homocysteine-homocysteine) and as the mixed disulphide homocysteine-cysteine by amino acid analysis[2].

In the plasma of normal fasting man, the presence of homocysteine-cysteine mixed disulphide only recently has been demonstrated by Gupta and Wilcken[3]; the women show significantly lower plasma levels than men[4]. We were able to confirm these findings[5]. In patients with homocystinuria, precocious arteriosclerosis and serious thromboembolic events are the most life-threatening complications. Moreover, in animal experiments, chronic homocystinemia has been shown to produce endothelial damage and arteriosclerosis[6,7,8] whereas, in cultures of human endothelial cells, addition of homocysteine and of homocystine to the medium causes cell injury that is considered to play a significant role in the genesis of arteriosclerosis[9,10]. These data led Gupta and Wilcken to suggest that the difference in plasma concentrations of the mixed disulphide between men and women, albeit small, could contribute over the years to the so far unexplained sex difference in the prevalence of occlusive vascular disease.

To further explore sex differences in methionine handling between men and women in different age classes, we studied the capacity to transsulphurate methionine in pre- and postmenopausal women and in groups of younger and older men of comparable age. Therefore, we loaded the subjects orally with methionine and measured during 8 h thereafter the serum profiles of methionine, homocystine, and the homocysteine-cysteine mixed disulphide.

## METHODS

Amino acid measurements before and after L-methionine loading were made in four groups of healthy subjects: (*a*) 10 older men aged 45-61 yr (53 ± 6, mean ± SD), (*b*) 10 younger men aged 22-35 yr (30 ±3, mean ± SD), (*c*) 10 women aged 45-59 yr (54 ± 3, mean ± SD) who all were postmenopausal for at least 0.5-10 yr (median 5). Their gonadotropins (luteinizing hormone 20-100 IU/liter, follicle-stimulating hormone, 30-120 IU/liter) and estradiol levels (<0.1 nmol/liter) were without exception in the postmenopausal ranges. (*d*) 10 young women aged 14-42 yr (25 ± 9, mean ± SD) who all had regular menses and did not take oral contraceptives. Neither between the groups of postmenopausal women and older men nor between the groups of young women and young men was the age difference significant ($P > 0.10$). None of the subjects took any drug or vitamin supplements. They all had normal renal function (serum urea, 3.0-7.0 mmol/liter; creatinine, 60-100 µmol/liter), normal liver function (serum glutamic pyruvic transaminase and serum glutamic oxalacetic transaminase <25 U/liter), and normal plasma levels of vitamin $B_6$ (44-96 nmol/liter) and folic acid (8-30 nmol/liter).

After an overnight fast, an oral L-methionine load was administered at 9.00 a.m. in a dose of 0.1 g (0.7 mmol)/kg body weight. From the height and the weight of the individual subjects, body surface was derived by using the nomogram based on the formula of Du Bois and Du Bois[11]. The mean methionine loads in the respective groups were 4.17 ± 0.09 SEM, 3.89 ± 0.07, 3.78 ± 0.10, and 3.71 ± 0.15 g/m$^2$ body surface. During the test, they stayed in the metabolic ward and used a methionine-free breakfast and luncheon with a total content of 14 mg methionine in 2 g of proteins[12], 95 g of carbohydrates, and 31 g of fats (670 cal, 2,814 J). Informed consent was obtained from all volunteering subjects after approval of the protocol by the hospital's Ethical Committee.

Methionine, homocystine, and the homocysteine-cysteine mixed disulphide in the serum were measured by ion-exchange chromatography (LC 2000 amino acid analyser, Biotronik Wissenschaftliche Geräte, Munich) in venous blood samples taken at 9.00 a.m. immediately before and 1, 2, 3, 4, 6, and 8 h after methionine loading. Metabolic clearances were calculated from the formula by which doses methionine administered (µmol/kg body weight) were divided by the area under the disappearance curves (µmol/liter × hour) after fitting the serum methionine data to a bi-exponential equation by using the PHARMFIT computer program[13]. The serum for amino acid analysis was deproteinized within 10 min after sampling by addition of equal volumes of 12.5% sulphosalicylic acid. Gamma-amino-butyric acid was used as an external standard. The deproteinized serum was stored at -20°C until analysis. All samples from one individual were measured in the same run within 4 wk to avoid systematic decreases in levels of sulphur-containing amino acids, which are observed after storage periods longer than 2 mo. Trace amounts of homocystine and homocysteine-cysteine mixed disulphide <0.1 µmol/liter are denoted as not detectable (ND) and quantitated as 0 in calculations of means.

Statistical analysis was performed by using Wilcoxon's two sample test (*P*) and Wilcoxon's paired rank test ($P^{\circ}$). Mean values ± 1 SEM are given.

RESULTS

Serum amino acid levels in the fasting state and maximal concentrations after methionine loading (Figs. 1 and 2, Table 1) were studied.

*Older and younger men.* The individual fasting methionine levels in the older men ranged from 14 to 41 µmol/liter and in the younger men from 16 to 47 µmol/liter. The mean fasting methionine levels (27 ± 2 vs. 30 ± 3 µmol/liter) did not differ between both groups, neither did the mean maximal concentration after loading. The individual peak levels after loading ranged from 871 to 1,801 µmol/liter in the older group and from 623 to 1,487 µmol/liter in the younger group. From the individual postload curves, we calculated a mean clearance of 0.10 ± 0.03 liter/h per kilogram for the older men and 0.07 ± 0.02 liter/h per kilogram for the younger men ($P$>0.10). Homocystine was not present in measurable amounts in any of the 20 men in the fasting state whereas, after methionine loading, homocystine levels rose by detectable concentrations in 12 out of these 20 men to a maximum of 0.4-1.9 µmol/liter. The mean homocysteine-cysteine mixed disulphide levels in the fasting state (3.1 ± 0.3 vs. 3.5 ± 0.5 µmol/liter) were similar in both groups; they ranged from 1.9 to 4.4 µmol/liter in the older men and from 1.1 to 5.6 µmol/liter in the younger men. After loading, the disulphide levels rose significantly in all of them ($P$°<0.01) to maximal concentrations that ranged from 8.4 to 16.0 µmol/liter in the older men and from 7.8 to 16.2 µmol/liter in the younger men. The mean levels in both groups increased by 8.5 ± 0.8 µmol/liter (321 ± 46%) in the older group and by 9.0 ± 0.7 µmol/liter (326 ± 63%) in the younger group ($P$>0.10).

*Postmenopausal women.* The fasting methionine levels ranged from 16 to 27 µmol/liter, whereas the mean concentration in postmenopausal women (21 ± 1 µmol/liter) was slightly lower than in both groups of men ($P$<0.05). The peak levels after loading ranged from 774 to as high as 1,406 µmol/liter. In contrast to the fasting levels, the mean peak levels did not differ from those in men. A clearance of methionine of 0.08 ± 0.02 liter/h per kilogram was calculated for this group. As in men, fasting homocystine levels were not detectable in these postmenopausal women. After loading with methionine, homocystinemia occurred in each of the women with maximal levels ranging from 1.3 to 4.6 µmol/liter. The mean homocystine level in these women (2.9 ± 0.3 µmol/liter) was significantly higher than in both groups of men ($P$<0.01). The fasting homocysteine-cysteine mixed disulphide concentrations in the postmenopausal women ranged from 1.3 to as high as 4.9 µmol/liter, and their mean (2.6 ± 0.4 µmol/liter) did not differ significantly from that in both groups of men. The peak levels achieved after methionine loading ranged from 10.3 to 25.7 µmol/liter and the mean maximal level (18.4 ± 1.5 µmol/liter) was significantly higher than in both groups of men ($P$<0.01).

*Table 1.* Means (±SEM) and ranges of serum amino acid levels in the fasting state and the maximal concentrations after methionine loading in normal men and
in normal post- and premenopausal women

| | Age | Methionine | | Homocystine | | Homocysteine-cysteine Mixed Disulphide | |
|---|---|---|---|---|---|---|---|
| | yr | Fasting | After load* | Fasting | After load µmol/liter | Fasting | After load |
| Men | | | | | | | |
| Older men | 53± 2 | 27± 2 | 1132± 80 | ND† | 0.6±0.2 | 3.1±0.3 | 11.6± 0.7 |
| (n=10) | 45 -61 | 14 -41 | 871 -1801 | | N.D. (5)§-1.9 | 1.9 -4.4 | 0.4 -16.0 |
| Young men | 30± 1 | 30± 3 | 993± 92 | ND | 0.8±0.2 | 3.5±0.5 | 12.5± 0.8 |
| (n=10) | 22 -35 | 16 -47 | 623 -1487 | | N.D.(3) -1.7 | 1.1 -5.6 | 7.8 -16.2 |
| Postmenopausal women | 54± 1 | 21± 1 | 1107± 53 | ND | 2.9±0.3 | 2.6±0.4 | 18.4± 1.5 |
| (n=10) | 45 -59 | 16 -27 | 774 -1406 | | 1.3 -4.6 | 1.3 -4.9 | 10.3 -25.7 |
| Premenopausal women | 25± 3 | 26± 2 | 1033± 60 | ND | ND | 0.9±0.3 | 6.8± 0.4 |
| (n=10) | 14 -42 | 16 -38 | 760 -1376 | | | 0.2 -2.6 | 5.0 - 9.1 |

* Maximal levels achieved 1-8 h after methionine loading (0.1 g/kg).
† Not detectable.
§ Number of patients indicated in parentheses.

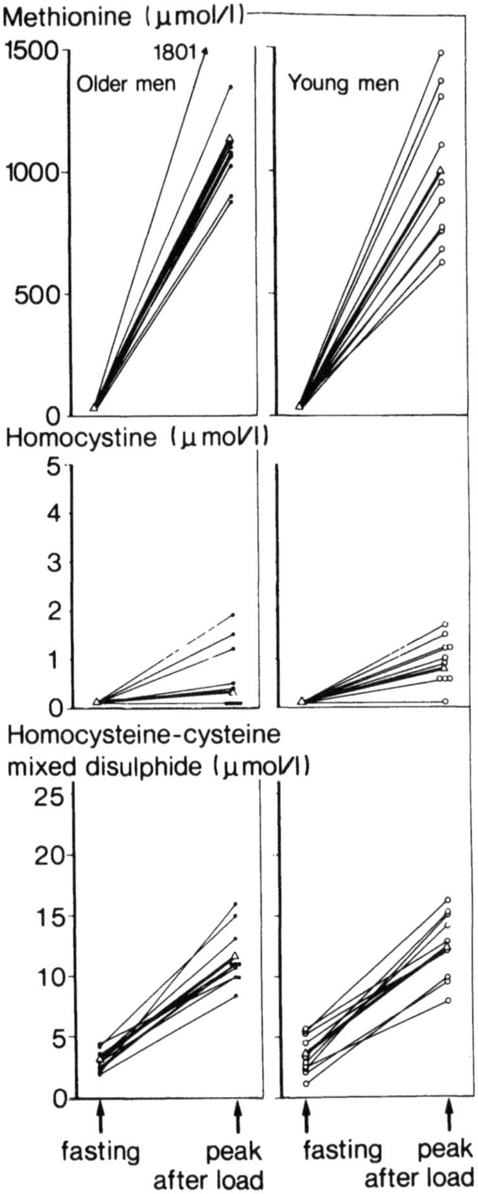


*Fig. 1.* Individual and mean (heavy bars) serum amino acid levels in the fasting state and peak concentrations after methionine loading (given in micromoles/liter) in 10 normal older men (●——●) and 10 normal young men (○——○). Concentrations <0.1 μmol/liter are denoted as nondetectable.

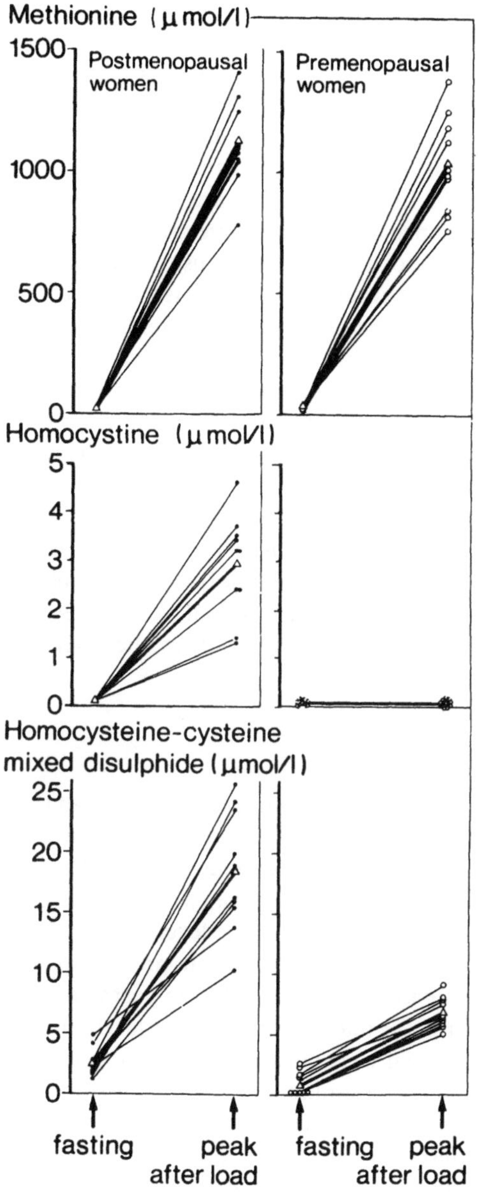


*Fig. 2.* Individual and mean (heavy bars) serum amino acid levels in the fasting state and peak concentrations after methionine loading (given in micromoles/liter) in 10 normal postmenopausal women (●——●) and 10 normal premenopausal women (○——○). Concentrations <0.1 µmol/liter are denoted as nondetectable.

*Premenopausal women.* The mean fasting methionine level (26 ± 2 μmol/liter) in the premenopausal women was slightly higher than in the postmenopausal group ($P<0.05$). The mean peak level after loading (range from 760 to 1,376 μmol/liter) did not differ from that in postmenopausal women. When compared with both groups of men, the mean fasting and the postloading methionine levels in the premenopausal women were virtually identical. The clearance of methionine, 0.08 ± 0.02 liter/h per kilogram, was not significantly different from the mean clearance in the other three groups ($P>0.10$). Homocystine was not detected in any of the premenopausal women in the fasting state as was assessed in the postmenopausal women and in both groups of men. In contrast to the homocystine profiles after methionine loading in the postmenopausal women and in both groups of men, homocystinemia did not occur in any of these premenopausal women. Furthermore, the mean fasting disulphide level (0.9 ± 0.3 μmol/liter) in the premenopausal group was at least three times lower than the mean value in any of the other groups ($P<0.02$-$0.01$). After methionine loading, individual peak disulphide increments in the premenopausal women (range 4.2-7.4 μmol/liter, mean 5.9 ± 0.8 μmol/liter) were without exception lower than in the postmenopausal women (range 7.9-22.9 μmol/liter, mean 15.8 ± 4.9 μmol/liter, $P<0.00025$). The mean peak disulphide increment in the premenopausal women was also significantly lower than that in both groups of men ($P<0.02$ vs. the older men and $P<0.005$ vs. the younger men).

## DISCUSSION

The present study shows that in men fasting methionine levels are higher than in women as reported earlier (5, 14, 15). However, our data demonstrate that this difference only holds for women in the postmenopausal state. No such difference was found between men and premenopausal women. Our data contradict the earlier statement that fasting methionine levels do not differ between pre- and postmenopausal women[16]. Indeed, in premenopausal women, these levels appeared to be significantly higher than in postmenopausal women. After methionine loading, the maximal methionine levels and the clearances in the premenopausal women were comparable with those in the other three groups. These results indicate a similar load to the transsulphuration pathway in the four groups. The present data illustrate that methionine loading uncovers a so far unreported difference in the occurrence of homocystinemia between pre- and postmenopausal women. Furthermore, the data in both groups of men strongly suggest that the unique absence of homocystinemia after loading in premenopausal women is not due to their young age. As first suggested by Mudd and Poole[17] and afterwards demonstrated by others[4,5], the mean fasting homocysteine-cysteine mixed disulphide levels in men are higher than in women. Our study clearly illustrates that this observation only holds true for premenopausal women, whereas these levels in postmenopausal women were not discernible from those in young and older men. It is intriguing that, in the fasting state, mean levels of the disulphide in the premenopausal women were on the average three times lower than in postmenopausal women whereas no homocystine was found in both groups. This unique characteristic of methionine handling in premenopausal women is further illustrated by the significantly lower increment of the disulphide despite a similar methionine load (also, after correction of the dose per kilogram for body surface) at equally high serum levels of this amino acid thereafter. Comparing the increments of the disulphide levels in the men and women of both age groups, the conclusion seems inevitable that age per se cannot be the factor responsible for the observed difference between pre- and postmenopausal women.

To summarize, the unique features of methionine handling in the premenopausal women are: (*a*) higher fasting methionine levels than in the postmenopausal women, *b*) no homocystinemia after loading with methionine in contrast to its invariable presence in the postmenopausal ones, and (*c*) much lower disulphide levels both before and after methionine loading. Although it would be attractive to hypothesize from these findings that premenopausal women better remethylate homocysteine to methionine, our data do not allow this thesis. Differences in transsulphuration rate or in renal excretion of homocysteine might be alternative explanations.

The methionine handling in premenopausal women is not only unique in comparison with postmenopausal women but also in comparison with young men; the latter showed significantly higher homocysteine-cysteine mixed disulphide levels and, in the majority of them, overt homocystinemia after methionine loading. When considered together, these observations indicate that premenopausal women are better protected against accumulation of homocysteine than young men.

Our data illustrate differences in methionine handling also between the postmenopausal women and the older men. They suggest that homocysteine accumulates much more in older women than in older men. This difference is the more convincing when the slightly higher load of methionine (absolute and per square meter body surface) in the older men is taken into account. This finding has not been reported so far and again accentuates that pre- and postmenopausal women differ markedly in their handling of methionine, and thus are unlike young and older men who, according to the data in Table 1, did not show differences in any of the variables studied.

Our findings finally indicate that a standardized methionine load can lead to overt homocystinemia in normal men and normal postmenopausal women and that an increase in the mixed disulphide levels even up to those found in obligate heterozygotes[18] in normal postmenopausal women could give rise to the erroneous conclusion of the presence of heterozygous classical homocystinuria.

Wilcken and Gupta suggested that the slightly but significantly lower fasting plasma levels of the homocysteine-cysteine mixed disulphide in women in comparison with men might be a contributing factor to the unexplained reduced proneness of women to develop vascular disease[4]. Probably most, if not all, women studied by the latter authors were in the premenopausal state as we deduced from the young age of their group. The observation by the same authors of accumulation of homocysteine in patients with chronic renal failure[19,20,21] and in young patients with angiographically established coronary artery disease[22] gives further support to the hypothesis that a reduced capacity to metabolize methionine is a risk factor for the development of premature arteriosclerosis.

The data in our study of normal men and women confirm that premenopausal women have lower fasting amounts of homocysteine-cysteine mixed disulphide than men. In addition, our data establish that such a sex difference does not exist between postmenopausal women and older men of similar age. After ingestion of a major methionine load to stress the transsulphuration pathway, premenopausal women do not exhibit homocystinemia and show a significantly lesser rise of homocysteine-cysteine mixed disulphide than postmenopausal women and younger and older men. A number of papers have demonstrated a lower rate of coronary artery disease in premenopausal women as compared with men of the same age, and an increasing incidence of cardiovascular disease in women after menopause is completed naturally or by castration[23-30]. Based on our data, we speculate that the unique efficiency in methionine handling in premenopausal women, which leads to a virtual absence of circulating homocystine and much lower levels of the disulphide, might be a natural defense mechanism to protect women in the reproductive life cycle against vascular damage.

REFERENCES

1. Mudd SH and Levy HL. 1983. Disorders of transsulphuration. In: Metabolic Bases of Inherited Disease. Stanbury JB, editor. McGraw-Hill, Inc., New York, Fifth ed. 522–59.
2. Perry TL, 1974. Homocystinuria. In: Heritable Disorders of Amino Acid Metabolism. Nyan WL, editor. John Wiley & Sons, Inc., New York. 395–428.
3. Gupta VJ and Wilcken DEL, 1978. The detection of cysteine-homocysteine mixed disulphide in plasma of normal fasting man. Eur J Clin Invest 8: 205–7.
4. Wilcken DEL and Gupta VJ, 1979. Cysteine-homocysteine mixed disulphide: differing plasma concentrations in normal men and women. Clin Sci (Lond.). 57: 211–5.
5. Boers GHJ, Smals AGH, Drayer JIM, Trijbels FJM, Leermakers AI and Kloppenborg PWC, 1983. Pyridoxine treatment does not prevent homocystinemia after methionine loading in adult homocystinuria patients. Metab Clin Exp 32: 390–7.
6. Harker LA, Slichter SJ, Scott CR and Ross R, 1974. Homocystinemia. Vascular injury and arterial thrombosis. N Engl J Med 291: 537–43.
7. Mc Cully KS and Wilson RB, 1975. Homocysteine theory of arteriosclerosis. Atherosclerosis 22: 215–27.

8. Harker LA, Ross R, Slichter SJ and Scott CR, 1976. Homocystine-induced arteriosclerosis. The role of endothelial cell injury and platelet response in its genesis. J Clin Invest 58: 731–41.
9. Wall RT, Harlan JM, Harker LA and Striker GE, 1980. Homocysteine-induced endothelial cell injury in vitro: a model for the study of vascular injury. Thromb Res 18: 113–21.
10. de Groot PG, Willems Ch, Boers GHJ, Gonsalves MD, van Aken WG and van Mourik JA, 1983. Endothelial cell dysfunction in homocystinuria. Eur J Clin Invest 13: 405–10.
11. Du Bois D, and Du Bois E, 1916. Formula to estimate the approximate surface area if height and weight known. Arch Intern Med 17: 863–71.
12. Food Policy and Food Science Service, Nutrition Division, Food and Agriculture Organization, Rome, 1970. Amino acid content of foods and biological data on proteins. 36–162.
13. Bozler G and van Rossum JM, 1982. Pharmacokinetics during drug development: data analysis and evaluation techniques. Gustav Fischer Verlag, New York. 210.
14. Armstrong MD and Stave U, 1973. A study of plasma free amino acid levels. II. Normal values for children and adults. Metab Clin Exp 22: 561–9.
15. Oepen H and Oepen I, 1965. Geschlechtsspezifische Konzentrationsunterschiede der Serum-Aminosäuren. Klin Wochenschr 43: 211–4.
16. Milsom JP, Morgan MY and Sherlock S, 1979. Factors affecting plasma amino acid concentrations in control subjects. Metab Clin Exp 28: 313–9.
17. Mudd SH and Poole JR, 1975. Labile methyl balances for normal humans on various dietary regimens. Metab Clin Exp 24: 721–35.
18. Sardharwalla IB, Fowler B, Robins AJ and Komrower GM, 1974. Detection of heterozygotes for homocystinuria. Arch Dis Child 49: 553–9.
19. Wilcken DEL and Gupta VJ, 1979. Sulphur containing amino acids in renal failure with particular reference to homocystine and cysteine-homocysteine mixed disulphide. Eur J Clin Invest 9: 301–7.
20. Wilcken DEL, Gupta VJ and Reddy SG, 1980. Accumulation of sulphur-containing amino acids including cysteine-homocysteine in patients on maintenance haemodialysis. Clin Sci (Lond.) 58: 427–30.
21. Wilcken DEL, Gupta VJ and Betts AK, 1981. Homocysteine in the plasma of renal transplant recipients: effect of cofactors for methionine metabolism. Clin Sci (Lond.) 61: 743–9.
22. Wilcken DEL and Wilcken B, 1976. The pathogenesis of coronary artery disease. A possible role for methionine metabolism. J Clin Invest 57: 1079–82.
23. Ryan KJ, 1976. Estrogens and atherosclerosis. Clin Obstet Gynaecol 19: 805–15.
24. Gordon T, Kannel WB, Hjortland MC and McNamara PM, 1978. Menopause and coronary heart disease. The Framingham study. Ann Intern Med 89: 157–61.
25. Robinson RW, Higano N and Cohen WD, 1959. Increased incidence of coronary heart disease in women castrated prior to menopause. Arch Intern Med 104: 908–13.
26. Sznajderman M and Oliver MF, 1963. Spontaneous premature menopause, ischaemic heart disease and serum-lipids. Lancet I: 962–5.
27. Oliver MF and Boyd GS, 1959. Effects of bilateral ovariectomy on coronary-artery disease and serum-lipids levels. Lancet II: 690–4.
28. Bengtsson C, 1973. Ischemic heart disease in women. Act Med Scand (Suppl.) 549: 1–128.
29. Rosenberg L, Hennekens ChH, Rosner B, Belanger Ch, Rothman KJ and Speizer FE, 1981. Early menopause and the risk of myocardial infarction. Am J Obstet Gynecol 139: 47–51.
30. Centerwall BS, 1981. Premenopausal hysterectomy and cardiovascular disease. Am J Obstet Gynecol 139: 58–61.

*Chapter 7*

# Improved identification of heterozygotes for homocystinuria due to cystathionine synthase deficiency by the combination of methionine loading and enzyme determination in cultured fibroblasts

Godfried H.J. Boers[1], B. Fowler[2], Anthony G.H. Smals[1], Frans J.M. Trijbels[3], Annelies I. Leermakers[1], Wim J. Kleijer[4], and P.W.C. Kloppenborg[1]

[1] *Department of Medicine, Division of Endocrinology, St. Radboud Hospital, University of Nijmegen, The Netherlands*

[2] *Willink Biochemical Genetics Unit, Royal Manchester Children's Hospital, Pendlebury, Manchester, UK*

[3] *Department of Pediatrics, St. Radboud Hospital, University of Nijmegen, The Netherlands*

[4] *Department of Clinical Genetics, Erasmus University, Rotterdam, The Netherlands*

SUMMARY

Previous data on tentative identification of the carrier state for homocystinuria due to cystathionine synthase deficiency using methionine loading or measurement of cystathionine synthase activity in tissue extracts are conflicting. We studied the results of standardized oral methionine loading in 20 obligate heterozygotes and compared them with those of determination of cystathionine synthase activity in cultured fibroblasts. Special attention was devoted to our recently reported observation on the small but striking differences in methionine metabolism between healthy pre- and postmenopausal women and men. Fasting and after load peak levels of methionine in serum did not discriminate the carriers from the control subjects. The mean fasting level of total homocysteine was only significantly higher in the group of premenopausal heterozygotes than in the corresponding control group. Nevertheless, the individual values overlapped with the normal range in 4 of 12 premenopausal heterozygotes. After loading peak levels of total homocysteine in 18 out of the 20 obligate heterozygotes exceeded the upper limit of the ranges in the three control groups. Thus, this parameter discriminated 90% of the obligate carriers. Measurement of cystathionine synthase activity in cultured fibroblasts from a skin biopsy identified the obligate heterozygotes to a similar degree (85%). No significant correlation between the measurements of cystathionine synthase activity and the after load peak levels of total

*Hum Genet 1985; 69: 164-9*

homocysteine in the individual heterozygotes was established. Combination of both methionine loading and determination of cystathionine synthase activity in cultured fibroblasts identified all of these carriers.

## INTRODUCTION

Homocystinuria due to cystathionine synthase (CS) deficiency, an inborn error of transsulphuration of the essential amino acid methionine, is inherited as an autosomal recessive trait. Data derived from screening programs to detect homozygous homocystinurics in newborns from various populations showed that the frequency of heterozygosity for this condition can be as high as 0.5 to 1.5% (Mudd and Levy 1983). In the homozygous state elevated levels of methionine, homocystine, and its mixed disulphide homocysteine-cysteine, in combination with decreased concentrations of cystine in the blood and urine are pathognomonic. In these patients CS activity in cultured fibroblasts is reportedly less than 10% of the mean control activity or not detectable (Mudd and Levy 1983; Boers et al. 1983a). After the recognition of the syndrome in 1962, attempts have been made to develop a reliable test for the identification of the heterozygous state. Based upon the finding that heterozygotes showed nearly 50% of the normal CS activity in liver tissue (Finkelstein et al. 1964), it was a logical approach to investigate whether loading by methionine and measuring sulphur-containing metabolites in blood or urine thereafter, could discriminate heterozygotes from normal subjects. Early studies in a small number of obligate heterozygotes failed to do so (White et al. 1964; Brenton et al. 1965; Kennedy et al. 1965; Laster et al. 1965a; Dunn et al. 1966; Chase et al. 1967). In 1974 Sardharwalla et al. (1974) distinguished 12 obligate heterozygotes from 12 control subjects using standardized methionine loading. Up to 24 h after the loading the heterozygotes had detectable levels of homocystine in plasma in contrast to the controls and significantly higher levels of the disulphide. Furthermore, the ratios of homocystine: cystine and of homocysteine-cysteine: cystine concentrations in the urine 24 h after loading clearly differentiated the two groups. Very recently, Wilcken et al. (1983) also demonstrated a statistically significant difference in plasma concentrations of homocystine and homocysteine-cysteine disulphide after similar methionine loading between a group of 17 obligate heterozygotes and a group of 17 normal controls, although there was a considerable overlap between the individual levels in 10 of them. We showed that pre- and postmenopausal women and men differ in their handling of methionine (Boers et al. 1983b). In that study we indicated that this difference might interfere with the detection of heterozygotes for homocystinuria especially in postmenopausal women. Taking into account the remarkable differences in increases of homocystine and homocysteine-cysteine after methionine loading between pre- and postmenopausal women and men, we evaluated results of methionine loading in 20 obligate heterozygotes for homocystinuria.

A more direct approach to identify heterozygosity is the establishment of the deficiency of CS activity in tissue extracts. Results of measurements of CS activity in liver biopsy specimen have been reported so far in only 10 obligate heterozygotes (Finkelstein et al. 1964; Laster et al. 1965b; Longhi et al. 1977; Gaull et al. 1974). In the first report on the results of determination of CS activity in extracts of cultured skin fibroblasts in obligate heterozygotes this method appeared to be a reliable test for identification, as all values in the six obligate heterozygotes studied were below the normal range (Fleisher et al. 1973). However, in three later studies an overlap of the values of the enzyme activity with those in control subjects was found in 13 of 17 obligate heterozygotes (Uhlendorf et al. 1973; Bittles and Carson 1981 a,b). In this study we systematically compared the results of standardized methionine loading with the activity of CS in cultured fibroblasts in a group of 20 obligate heterozygous adults.

## MATERIALS AND METHODS

### *Obligate heterozygotes and control subjects*

Twenty obligate heterozygotes, parents or children of 13 patients suffering from homocystinuria due to cystathionine synthase deficiency, were studied. In these 13 patients from 9 sibships the homozygous state had been demonstrated by established hypermethioninemia, homocystinemia, elevated levels of homocysteine-cysteine mixed disulphide, and lowered cystine levels in fasting serum. In addition, determination of CS activity in cultured fibroblasts has shown non-detectable or residual enzyme activities less than 4% of the lowest value found in normal subjects. The data of 10 of these patients have been reported before (Boers et al. 1983a, 1984). Twelve of the 20 obligate heterozygotes were premenopausal women (aged 18 to 52 years; 35 ± 9, mean ± SD), two women were postmenopausal (57 and 61 years old), and six were men aged 30 to 61 years (46 ± 12, mean ± SD). The controls (Table 1) consisted of 40 healthy subjects: 10 premenopausal women, 10 older women who all were in the postmenopausal state for at least 0.5 to 10 years (median 5 years), 10 younger men and 10 older men (Boers et al. 1983b). None of the obligate heterozygotes and the control subjects took any drugs or vitamin supplements; the premenopausal women did not use oral contraceptives. All obligate heterozygotes and control subjects studied had normal renal function (serum urea, 3.0-7.0 mmol/l; creatinine, 60-100 µmol/l), normal liver tests (serum glutamic pyruvic transaminase and serum glutamic oxalacetic transaminase less than 25 U/l), and normal plasma levels of vitamin $B_6$ (44-96 nmol/l) and folic acid (8-30 nmol/l). Informed consent was obtained from all volunteering obligate heterozygotes and control subjects after approval of the protocol by the hospital's Ethical Committee.

## METHODS

After an overnight fast an oral L-methionine load was administered at 9.00 a.m. in a dose of 0.1 g (0.7 mmol)/kg body weight. During the test all subjects stayed in the metabolic ward and ate a methionine-restricted breakfast and luncheon with a total content of 14 mg methionine in 2 g of proteins, 95 g of carbohydrates, and 31 g of fats (670 Cal, 2.814 J).

Methionine, homocystine, and homocysteine-cysteine mixed disulphide were measured in the serum by ion-exchange chromatography (Biotronik LC 2000 amino acid analyser) in venous blood samples taken at 9.00 a.m., immediately before, and 1, 2, 3, 4, 6, and 8 h after methionine loading. Details of the analytical methods have been given recently elsewhere (Boers et al. 1983b). The total amount of circulating nonprotein-bound homocysteine was calculated as twice the concentration of homocystine (homocysteine-cysteine) plus the concentration of homocysteine-cysteine mixed disulphide, in μmol/l. The data of methionine loading in the four groups of control subjects have been reported before (Boers et al. 1983b), and the after load peak levels of total homocysteine in these controls have been calculated from the earlier published data of their after load peak levels of homocystine and homocysteine-cysteine mixed disulphide.

The determination of CS activity in fibroblast cultures from skin biopsies in the obligate heterozygotes was performed using a previously reported method (Fowler et al. 1978), with one modification: paper chromatography was performed with development times of at least 60 h to ensure the separation of $^{14}$C-cystathionine from an additional radioactive compound formed in the assay. Assays were performed on at least two separate batches of cells. The results of this assay in the obligate heterozygotes were compared with those established in 23 controls.

Statistical analyses were performed using Wilcoxon's two sample test (*P*) and Spearman's rank correlation test ($r_s$, $P^+$). Unless otherwise stated, mean values ± 1 SEM are given.

## RESULTS

### *Serum levels of methionine in the fasting state and maximal concentrations after methionine loading (Table 1)*

Fasting methionine levels in the group of premenopausal heterozygotes ranged from 22 to 34 μmol/l and they all were in the range of control premenopausal women. The mean fasting methionine level (27 ± 1μmol/l)

*Table 1.* Means (±SEM) and ranges of serum levels of methionine and total homocysteine in the fasting state and the maximal concentrations after methionine loading in 20 obligate heterozygotes (12 premenopausal and 2 postmenopausal women, and 6 men) and in 40 control subjects (10 premenopausal and 10 postmenopausal women, and 20 men)

| | Age | Methionine | | Total homocysteine** | |
|---|---|---|---|---|---|
| | (years) | Fasting µmol/l | After load* µmol/l | Fasting µmol/l | After load** µmol/l |
| Premenopausal women | | | | | |
| Obligate heterozygotes | 35± 3 | 27± 1 | 860± 50 | 4.2± 0.8 | 41.0± 7.1 |
| (n=12) | 18 -52 | 22 -34 | 519 -1068 | 1.4 -11.6 | 15.2 -99.8 |
| Control subjects | 25± 3 | 26± 2 | 1033± 60 | 0.9± 0.3 | 6.8± 0.4 |
| (n=10) | 14 -42 | 16 -38 | 760 -1376 | 0.2 - 2.6 | 5.0 - 9.1 |
| Postmenopausal women | | | | | |
| Obligate heterozygotes (n=2) | 57,61 | 31, 33 | 618, 1258 | 3.3, 3.6 | 20.6, 51.9 |
| Control subjects | 54± 1 | 21± 1 | 1107± 53 | 2.6± 0.4 | 24.1± 1.9 |
| (n=10) | 45 -59 | 16 -27 | 774 -1406 | 1.3 - 4.9 | 17.1 -34.9 |
| Men | | | | | |
| Obligate heterozygotes | 46± 5 | 32± 2 | 1154± 50 | 3.4± 0.8 | 26.7± 5.2 |
| (n=6) | 30 -61 | 24 -39 | 925 -1273 | 1.8 - 7.1 | 15.9 -51.5 |
| Control subjects | 42± 3 | 28± 2 | 1063± 65 | 3.3± 0.3 | 13.3± 0.8 |
| (n=20) | 22 -61 | 14 -47 | 623 -1801 | 1.1 - 5.6 | 7.8 -19.2 |

*Maximal levels achieved 1-8 h after methionine loading (0.1 g/kg)

**Calculated from 2 × concentration of homocystine plus concentration of homocysteine-cysteine mixed disulphide.

did not differ significantly from the mean in the controls (26 ± 2 μmol/l) ($P$>0.10). The two postmenopausal heterozygotes showed slightly higher methionine levels in the fasting state than control postmenopausal women. The fasting methionine levels in the male heterozygotes, ranging from 24 to 39 μmol/l, were all in the range of those in control men, and therefore the mean concentration (32 ± 2 μmol/l) was not significantly different from that in the controls (28 ± 2 μmol/l) ($P$>0.10).

The mean peak level of methionine after the loading in the group of heterozygous premenopausal women was slightly but not significantly lower than in the group of premenopausal controls (860 ± 50 μmol/l versus 1033 ± 60 μmol/l, $P$>0.10). None of these heterozygotes had peak levels of methionine exceeding the upper limit of the range in controls. The peak levels of methionine in the two postmenopausal heterozygotes were 618 and 1258 μmol/l, one below and the other in the range of postmenopausal controls. The after load peak levels of methionine in the heterozygous men ranged from 925 to 1273 μmol/l, and were without exception in the range of control men. Their mean did not differ significantly from the mean in the control group (1154 ± 50 μmol/l versus 1063 ± 65 μmol/l, $P$>0.10).

*Serum levels of total homocysteine in the fasting state and maximal concentrations after methionine loading (Table 1, Fig. 1)*

The mean fasting level of total homocysteine in the group of premenopausal

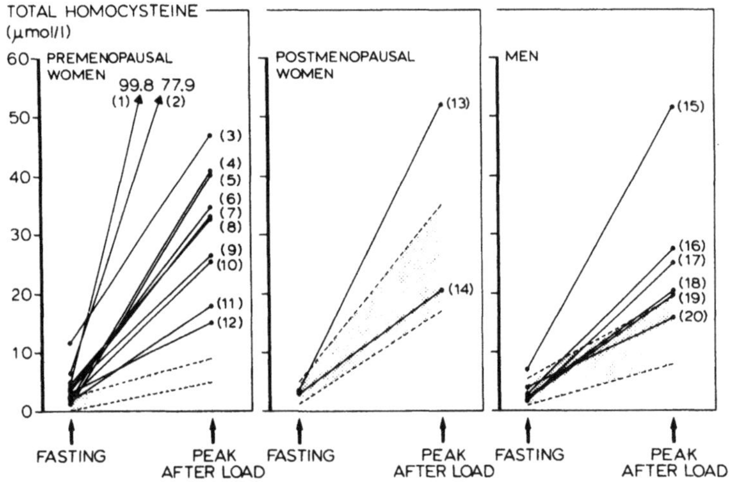


*Fig. 1.* Individual serum levels of total homocysteine in the fasting state and peak concentrations after methionine loading, in μmol/l, in 20 obligate heterozygotes (12 premenopausal and 2 postmenopausal women, and 6 men) (●———●) and the ranges of these levels in 40 control subjects (10 premenopausal and 10 postmenopausal women, and 20 men) (*shaded*). The numbers indicated in parentheses correspond with those of the individual obligate heterozygotes in Table 2.

heterozygotes (4.2 ± 0.8 μmol/l) was almost five-fold the mean concentration in the premenopausal controls (0.9 ± 0.3 μmol/l) ($P<0.01$). Nevertheless, in 4 out of 12 premenopausal heterozygotes the fasting levels were in the normal range. The two postmenopausal heterozygotes had fasting concentrations of total homocysteine of 3.3 and 3.6 μmol/l, both in the range of control postmenopausal women. The fasting levels in five of the six heterozygous men were also in the range of their controls.

Without exception the individual peak levels of total homocysteine achieved at 6 to 8 h after methionine loading in the group of premenopausal heterozygotes, ranging from 15.2 to 99.8 μmol/l, exceeded the range in control premenopausal women (Fig. 1). The mean level in the heterozygotes was about six-fold the mean in the premenopausal controls (41 ± 7.1 μmol/l versus 6.8 ± 0.4 μmol/l, $P<0.01$). From the postmenopausal heterozygotes one had an after load peak level of total homocysteine (20.6 μmol/l) in the range of the control postmenopausal women, whereas the other reached 51.9 μmol/l, well above the control upper limit. In the six heterozygous men peak levels of total homocysteine ranged from 15.9 to 51.5 μmol/l, in only one of them the peak value fell within the control range. The mean peak level of total homocysteine in the group of heterozygous men was twice the mean in the control group (26.7 ± 5.2 μmol/l versus 13.3 ± 0.8 μmol/l, $P<0.01$).

*Cystathionine synthase activities in cultured fibroblasts (Table 2, Fig. 2)*

CS activities measured in cultured fibroblasts without addition of pyridoxal-5′-phosphate (PLP) to the assay mixture ranged in the 20 obligate heterozygotes from 0.1 to 4.2 $nmol \cdot h^{-1} \cdot mg^{-1}$ and with 1 mM PLP added to the assay mixture from 0.1 to 5.4 $nmol \cdot h^{-1} \cdot mg^{-1}$. The activities measured both without and with added PLP in three obligate heterozygotes overlapped the lower end of the range obtained in a group of control subjects. Mean CS activities in the group of heterozygotes (without PLP 2.2 ± 0.2 $nmol \cdot h^{-1} \cdot mg^{-1}$, with PLP 3.0 ± 0.3 $nmol \cdot h^{-1} \cdot mg^{-1}$) were significantly lower than the respective mean activities in the control group ($P<0.01$). The correlation between the measured individual CS activities without or with added PLP and the after load levels of total homocysteine was not statistically significant ($r_s = -0.25$ and $r_s = -0.19$, N.S.). In the three obligate heterozygotes whose CS activities were within the lower limit of the control range (Nos. 11, 13, and 18), the total homocysteine levels after methionine loading were clearly higher than in the corresponding control groups (Table 2). In the two heterozygotes (Nos. 14 and 20) whose total homocysteine levels after methionine loading were within the corresponding normal ranges, the CS activities were 2.3 and 2.4 $nmol \cdot h^{-1} \cdot mg^{-1}$ and below the lower limit of the normal range.

*Table 2.* Cystathionine synthase activities measured in cultured fibroblasts, determined without and with addition of 1 mM pyridoxal-5'-phosphate (PLP) to the assay mixture, and the peak concentrations of total homocysteine after methionine loading in 20 obligate heterozygotes

| Obligate heterozygotes | Cystathionine synthase* $nmol \cdot h^{-1} \cdot mg^{-1}$ -PLP | +PLP | Peak total homocysteine µmol/l |
|---|---|---|---|
| 1 premenopausal women | 1.7 | 2.4 | 99.8 |
| 2 | 1.4 | 2.0 | 77.9 |
| 3 | 2.4 | 3.6 | 46.9 |
| 4 | 0.1 | 0.1 | 40.9 |
| 5 | 1.4 | 2.1 | 40.2 |
| 6 | 2.2 | 2.4 | 34.8 |
| 7 | 2.3 | 3.1 | 33.1 |
| 8 | 0.9 | 1.2 | 32.8 |
| 9 | 0.9 | 1.1 | 26.5 |
| 10 | 1.7 | 3.0 | 25.6 |
| 11 | 4.2 | 5.1 | 18.1 |
| 12 | 2.7 | 3.1 | 15.2 |
| 13 postmenopausal women | 3.7 | 4.8 | 51.9 |
| 14 | 2.3 | 3.1 | 20.6 |
| 15 men | 3.2 | 5.4 | 51.5 |
| 16 | 2.7 | 3.7 | 27.5 |
| 17 | 1.1 | 1.4 | 25.1 |
| 18 | 4.1 | 5.3 | 20.4 |
| 19 | 2.1 | 4.6 | 19.5 |
| 20 | 2.4 | 3.0 | 15.9 |
| Mean | 2.2 | 3.0 | |
| ±SEM | ±0.2 | ±0.3 | |
| Range | 0.1– 4.2 | 0.1– 5.4 | |
| Controls (n=23) | | | |
| Mean | 11.3 | 14.2 | |
| ±SEM | ±1.1 | ±1.5 | |
| Range | 3.6–25.7 | 4.9–32.1 | |

*Values are averages of at least two determinations

## DISCUSSION

The present study clearly shows that neither the fasting levels nor the peak concentrations of methionine after methionine loading reliably identify heterozygotes for homocystinuria, confirming the findings in some earlier reports (Brenton et al. 1965; Dunn et al. 1966; Sardharwalla et al. 1974; Wilcken et al. 1983) and in contrast with the data in one other report (Chase et al. 1967).

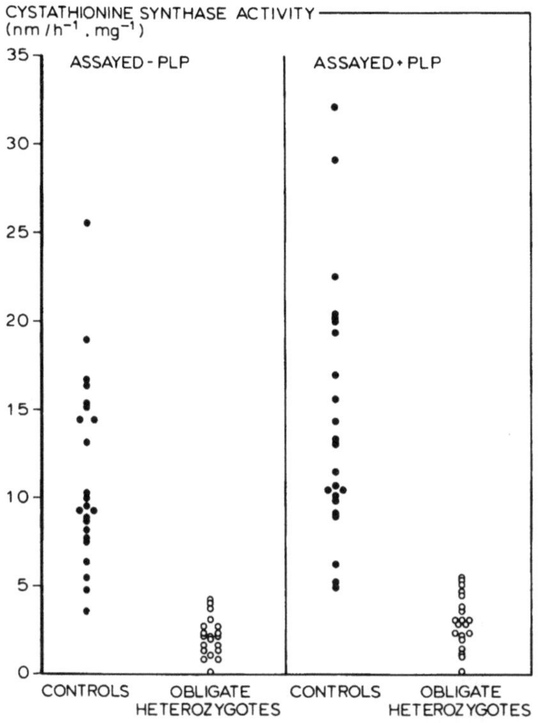


*Fig. 2.* Cystathionine synthase activities in nmol·h$^{-1}$·mg$^{-1}$ measured in cultured fibroblasts in 23 controls (●) and in 20 obligate heterozygotes (○), determined without (*left panel*) and with addition of pyridoxal-5′-phosphate (PLP), 1 mM, to the assay mixture (*right panel*).

This observation was not unexpected as we earlier demonstrated that even *homozygous* patients could not be detected merely on the strength of methionine levels in the blood after methionine loading (Boers et al. 1983a).

As far as total homocysteine is concerned, Sardharwalla et al. (1974) could neither detect homocystine nor homocysteine-cysteine mixed disulphide in the fasting state in 12 control subjects and in 7 out of 12 obligate heterozygotes, whereas Wilcken et al. (1983) reported lack of statistically significant difference in the mean fasting total homocysteine level between a group of 17 obligate heterozygotes and a group of 17 controls. The data in our study illustrate that only in the group of premenopausal heterozygotes the mean fasting level of total homocysteine is significantly higher than in control premenopausal women, but individual values showed overlap with the normal range in 4 out of 12.

Taking into account the differences in methionine metabolism between pre- and postmenopausal women and men (Boers et al. 1983b), standardized methionine loading identified 18 out of 20 obligate heterozygotes by peak total homocysteine levels clearly beyond the upper limit of the respective range in the three diverse control groups. In the 12 premenopausal heterozygous women the peak homocysteine levels discriminated them in all. When not differentiating the controls and the obligate heterozygotes according to sex and menopausal status, 13 out of 20 heterozygotes had peak levels of total homocysteine after methionine loading within the overall range of these concentrations in all 40 control subjects (Fig. 1). Even worse, when not accounting for the differences in methionine handling after loading before and after menopause, all 10 normal postmenopausal women would have been erroneously classified as heterozygotes for homocystinuria (Boers et al. 1983c). This may be a significant factor in the results of Wilcken et al. (1983) who reported that in 10 of 17 obligate heterozygotes total homocysteine levels at 12 h after methionine loading were within the range of control subjects, and this author did not discriminate the control subjects according to sex and menopausal state. Sardharwalla et al. (1974) found clear differentiation of all 12 methionine-loaded heterozygotes from controls, but it has to be stressed that these investigators could not detect measurable amounts of homocystine after methionine loading in any of their control subjects. This probably reflects the lower sensitivity of the amino acid analysis system used.

The more sophisticated approach by measuring CS activity in cultured skin fibroblasts failed to decisively discriminate heterozygotes from control subjects in earlier studies (Fleisher et al 1973; Uhlendorf et al. 1973; Bittles and Carson 1981a, b) as in only 10 of a total of 23 reported obligate heterozygotes the measured enzyme activity was below the range in controls. In our hands however, this procedure did identify 17 out of 20 obligate heterozygotes (85%). Nevertheless, this relatively high score proved not to be superior to that of the more simple standardized methionine loading test which discriminated 90% of the obligate heterozygotes, under the proviso of differentiating control subjects according to sex and menopausal state. When taking the results of both procedures together, all of our 20 obligate heterozygotes were decisively identified.

A significant correlation between the individual results with the two tests was not found. This is not altogether surprising since both homocysteine levels in serum after oral methionine loading and CS activity expression in fibroblasts are removed somewhat from the primary factor, i.e., liver CS activity, and are subject to non-genetic or exogenous influences. This is in spite of rigorous attempts to standardize both cell culture and CS assay conditions as well as the methionine loading test. Importantly, however, this study does provide an enzymatic basis for the observed metabolic differences between heterozygotes and control subjects.

Attention must be brought to heterozygote No. 4 who had surprisingly low CS activity in her fibroblasts, in fact in the range of homozygous patients, whereas the peak level of total homocysteine in her serum after methionine loading was clearly lower than those observed in *homozygous* patients (range 112-277 μmol/l, Boers et al. 1983a). Against this subject being homozygous is furthermore the fact that her mother is undoubtedly identified as a homozygote by both the methionine loading test and CS activity determination, whereas her father and her daughter showed normal total homocysteine levels after methionine loading. This emphasizes the possibility of influence of non-genetic factors on CS activity expression in fibroblasts and highlights the need to exercise caution and to consider all factors in interpretation of such results.

In summary, our study illustrates that both the standardized methionine loading test and CS activity measurement in cultured fibroblasts from skin biopsies are to a similar degree highly discriminatory in the identification of heterozygotes for homocystinuria due to CS deficiency. The combination of both tests identified all 20 carriers for this condition examined in this study.

## REFERENCES


Bittles AH, Carson NAJ (1981a) Homocystinuria: studies on cystathionine β-synthase, S-adenosylmethionine synthetase and cystathionase activities in skin fibroblasts. J Inherited Metab Dis 4: 3–6.

Bittles AH, Carson NAJ (1981b) Homocystinuria: the effect of pyridoxine supplementation on cultured skin fibroblasts. J Inherited Metab Dis 4: 7–9.

Boers GHJ, Smals AGH, Drayer JIM, Trijbels FJM, Leermakers AI, Kloppenborg PWC (1983a) Pyridoxine treatment does not prevent homocystinemia after methionine loading in adult homocystinuria patients. Metabolism 32: 390–397.

Boers GHJ, Smals AG, Trijbels FJ, Leermakers AI, Kloppenborg PW (1983b) Unique efficiency of methionine metabolism in premenopausal women may protect against vascular disease in the reproductive years. J Clin Invest 72: 1971–1976.

Boers GHJ, Trijbels JM, Smals AG, Kloppenborg PW (1983c) Methionine-loading can lead to the erroneous conclusion of heterozygosity for homocystinuria in normal postmenopausal women. Clin Genet 24: 289–290.

Boers GHJ, Polder TW, Cruysberg JRM, Schoonderwaldt HC, Peetoom JJ, van Ruyven TWJ, Smals AGH, Kloppenborg PWC (1984) Homocystinuria versus Marfan's syndrome: the therapeutic relevance of the differential diagnosis. Neth J Med 27: 206–212.

Brenton DP, Cusworth DC, Gaull GE (1965) Homocystinuria: metabolic studies on three patients. J Pediatr 67: 58–68.

Chase HP, Goodman SI, O'Brien D (1967) Treatment of homocystinuria. Arch Dis Child 42: 514–520.

Dunn HG, Perry TL, Dolman CL (1966) Homocystinuria. Neurology 16: 407–420.

Finkelstein JD, Mudd SH, Irreverre F, Laster L (1964) Homocystinuria due to cystathionine synthetase deficiency: the mode of inheritance. Science 146: 785–787.

Fleisher LD, Tallan HH, Beratis NG, Hirschhorn K, Gaull GE (1973) Cystathionine synthase deficiency: heterozygote detection using cultured skin fibroblasts. Biochem Biophys Res Commun 55: 38–44.

Fowler B, Kraus J, Packman S, Rosenberg LE (1978) Homocystinuria. Evidence of three distinct classes of cystathionine β-synthase mutants in cultured fibroblasts. J Clin Invest 61: 645–653.

Gaull G, Sturman JA, Schaffner F (1974) Homocystinuria due to cystathionine synthase deficiency: enzymatic and ultrastructural studies. J Pediatr 84: 381–390.

Kennedy C, Shih VE, Rowland LP (1965) Homocystinuria: a report in two siblings. Pediatrics 36: 736–741.

Laster L, Mudd SH, Finkelstein JD, Irreverre F (1965a) Homocystinuria due to cystathionine synthase deficiency: the metabolism of L-methionine. J Clin Invest 44: 1708–1719.

Laster L, Spaeth GL, Mudd SH, Finkelstein JD (1965b) Homocystinuria due to cystathionine synthase deficiency. Ann Intern Med 63: 1117–1142.

Longhi RC, Fleisher LD, Tallan HH, Gaull GE (1977) Cystathionine β-synthase deficiency: a qualitative abnormality of the deficient enzyme modified by vitamin $B_6$-therapy. Pediatr Res 11: 100–103.

Mudd SH, Levy HL (1983) Disorders of transsulphuration. In: Stanbury JB, Wyngaarden JB, Fredrickson DS (eds) The metabolic basis of inherited disease. McGraw-Hill, New York, pp. 522–559.

Sardharwalla IB, Fowler B, Robins AJ, Komrower GM (1974) Detection of heterozygotes for homocystinuria. Arch Dis Child 49: 553–559.

Uhlendorf BW, Conerly EB, Mudd SH (1973) Homocystinuria: studies in tissue culture. Pediatr Res 7: 645–658.

White HH, Thompson HL, Rowland LP, Cohen D, Araki S (1964) Homocystinuria. Trans Am Neurol Assoc 89: 24–27.

Wilcken DEL, Reddy SG, Gupta VJ (1983) Homocysteinemia, ischemic heart disease, and the carrier state of homocystinuria. Metabolism 32: 363–370.

*Chapter 8*

# High frequency of heterozygosity for homocystinuria in premature occlusive peripheral and cerebral arterial disease

Godfried H.J. Boers, M.D., Antony G.H. Smals, M.D., Frans J.M. Trijbels, Ph.D., Brian Fowler, Ph.D., Jan A.J.M. Bakkeren, Ph.D., Henny C. Schoonderwaldt, M.D., Wim J. Kleijer, Ph.D., Peter W.C. Kloppenborg, M.D.

*From the Departments of Medicine, Pediatrics and Neurology, St. Radboud Hospital, University of Nijmegen, The Netherlands; The Willink Biochemical Genetics Unit, Royal Manchester Children's Hospital, United Kingdom; The Department of Clinical Genetics, Erasmus University, Rotterdam, The Netherlands.*

ABSTRACT

Premature arteriosclerosis and thromboembolic events are well-known complications of homozygous homocystinuria due to cystathionine synthase deficiency. It is unknown whether heterozygosity for homocystinuria predisposes to premature vascular disease. We explored the frequency of excessive homocysteine accumulation after standardized methionine loading, to a degree as observed in obligate carriers, in 75 patients presenting with clinical signs of ischemic disease under the age of 50, 25 with occlusive peripheral arterial disease, 25 with occlusive cerebrovascular pathology, and 25 with myocardial infarction. In 7 patients of each of the former two groups but in none of the latter group heterozygosity for homocystinuria was established based upon pathological homocysteinemia after methionine loading and substantiated by the measurement of cystathionine synthase deficiency in skin fibroblast cultures. Because the frequency of heterozygosity for homocystinuria in the normal population is 1 in 70 at the most, we conclude that heterozygosity for homocystinuria predisposes to the development of premature occlusive arterial disease leading to common clinical disorders such as intermittent claudication and renovascular hypertension, especially in women, and ischemic cerebrovascular pathology.

*Submitted for publication*

## INTRODUCTION

Homocystinuria due to cystathionine synthase deficiency in homozygous form is frequently associated with severe vascular disease already occurring in infancy and childhood. The pathological accumulation of homocysteine in tissues and blood is generally held to initiate premature vascular occlusive disease by causing endothelial cell damage[1]. From a recent survey of 629 patients by Mudd et al.[2] we estimated a probability of thromboembolic events occurring before the age of 20 of 30% and before the age of 40 years of about 60%.

Definite data on mild homocysteinemia as a risk factor of premature vascular damage in carriers for this condition is lacking and, in turn, it is unknown whether in groups of patients with premature arteriosclerosis the frequency of heterozygosity for homocystinuria is higher than the estimated 1:70 to 200 frequency in the normal population. Wilcken et al. could not confirm a predominance of heterozygotes for homocystinuria among patients with premature *coronary* vascular disease[3] thereby withdrawing their earlier reported finding of pathological homocysteine accumulation after methionine loading in 25 patients with angiographically proven ischemic heart disease under the age of 50 years[4]. Reports on the frequency of carriership for homocystinuria in patients suffering from occlusive *peripheral* arterial or *cerebrovascular* disease are not available.

Therefore we studied the frequency of excessive homocysteine accumulation after standardized methionine loading in 75 patients presenting with signs of ischemic disease under the age of 50 years, 25 of whom had occlusive *peripheral* arterial disease, 25 occlusive *cerebrovascular* disease, and 25 *myocardial* infarction. In all of them occlusive arterial lesions had been angiographically proven. Based upon abnormally high homocysteine accumulation after methionine loading, we diagnosed the presence of heterozygosity for homocystinuria in about 30% of the patients in the groups with occlusive *peripheral* arterial and with *cerebrovascular* disease. In these patients heterozygosity for homocystinuria was further substantiated by measuring cystathionine synthase activity in cultured fibroblasts from skin biopsies. In contrast, in the group of patients with *myocardial* infarction abnormal homocysteine accumulation was not found in any of them.

## METHODS

### *Patients*

In a prospective study all patients successively diagnosed since January 1980 as suffering from premature arteriosclerotic disease, 25 patients with occlusive peripheral arterial disease, 25 with occlusive cerebrovascular disease, and 25 with myocardial infarction, underwent oral methionine loading within four to six weeks after the diagnosis. All patients met the criteria determined beforehand: age younger than 50 years, absence of diabetes, hyperlipoproteinemia, high blood pressure unless of renovascular origin, and angiographically proven arteriosclerotic lesions. The group of patients suffering from occlusive peripheral arterial disease consisted of 11 men and 14 women (age 38 ± 9 years, mean ± S.D.). In the group of patients with occlusive cerebrovascular disease 12 were men and 13 women (age 41 ± 7 years). The group of patients suffering from myocardial infarction consisted of 22 men and 3 women (age 36 ± 6 years). At the time of the test all patients had normal renal and liver function, and normal plasma levels of vitamin B6.

*Methionine loading test*

Details on the standardization of the oral methionine loading test and the analytical methods used in the determination of amino acids, have been published recently by this laboratory[5]. In summary: after an overnight fast an oral L-methionine load was administered at 9.00 a.m. in a dose of 0.1 g (0.7 mmol)/kg. During the test the patients consumed a methionine-restricted diet with a total content of 14 mg of methionine. Methionine, homocystine, and homocysteine-cysteine mixed disulphide were measured in serum by ion-exchange chromatography (Biotronik LC 2000) in venous blood samples taken at 9.00 a.m., immediately before, and 1, 2, 3, 4, 6, and 8 hours after methionine loading. The total amount of circulating non-protein-bound homocysteine in a blood sample was calculated as twice the concentration of homocystine (homocysteine-homocysteine) plus the concentration of homocysteine-cysteine mixed disulphide, in micromoles per liter. The limits of detection of both homocystine and homocysteine-cysteine are 0.1 μmol/l.

*Identification of heterozygotes for homocystinuria due to cystathionine synthase deficiency*

In the interpretation of the data of the methionine loading test in the patients due attention was paid to the earlier reported significant differences in methionine handling between pre- and postmenopausal women and men[5]. Patients were considered to be heterozygotes for homocystinuria due to cystathionine synthase deficiency when peak levels of homocysteine after methionine loading clearly exceeded the ranges obtained in the respective groups of normal control subjects, with the proviso of the absence of conditions known to disturb methionine handling such as renal failure, liver cirrhosis, and vitamin B6 deficiency. The validity of this procedure to identify heterozygotes for homocystinuria has been amply documented in a recent study from this laboratory[6]. In the patients with excessive homocysteinemia after methionine loading determinations of cystathionine synthase activity in skin fibroblasts were performed using a previously reported method[7]. The results of the methionine loading test and the enzyme assays were compared to those obtained in 20 adult obligate heterozygotes[6] and in 15 adult homozygous homocystinuria patients whose data have been published previously in the majority of the cases[7-10].

*Statistical analyses*
Statistical analyses were performed by using Wilcoxon's two sample test (P), the chi square test ($P^{+}$) and Spearman's rank correlation test ($r_s$, $P^{++}$). Unless otherwise stated, mean values ± 1 S.E.M. are given.
Informed consent was obtained from all patients after approval of the protocol by the hospital's Ethical Committee.

RESULTS

*Clinical data*
In none of the 25 patients with *myocardial* infarction due to proven premature occlusive arterial disease was pathological accumulation of homocysteine after methionine loading found. In 7 out of 25 patients with either proven occlusive *peripheral* arterial disease or occlusive *cerebrovascular* disease such accumulation was established and these patients therefore were considered to be heterozygotes for homocystinuria due to cystathionine synthase deficiency. In tables 1 and 2 clinical data of these patients are summarized. In addition, the peak homocysteine levels after methionine loading are given and compared with those in the remaining patients with normal handling of methionine. All heterozygotes with occlusive peripheral arterial disease were women whereas from the remaining 18 patients in this group 11 were men ($P^{+} < 0.025$). In the majority of both groups intermittent claudication or renovascular hypertension was the predominant presenting clinical sign. Table 2 illustrates the data in the group of patients with angiographically proven occlusive cerebrovascular disease. The sex ratio in these heterozygotes was similar to that in the remaining 18 patients without abnormal methionine handling. In the majority of the patients in both groups persistent neurologic damage prevailed. Eighty percent of the heterozygous and 70% of the non-heterozygous patients had smoked daily more than 10 cigarettes for a long period. Forty percent of the female heterozygous and 60% of the female non-heterozygous patients had taken oral contraceptives for longer than one year.

*Biochemical data*
In none of the 75 patients with premature occlusive arterial disease did the peak level of methionine after loading exceed the range of these concentrations established in normal control subjects, and the mean did not differ significantly from the mean peak level in the controls (965 ± 20 µmol/l vs 1066 ± 38 µmol/l, $P > 0.10$), indicating a similar challenge to the transsulphuration pathway leading from methionine to homocysteine.

In table 3 relevant data of the methionine loading tests in the 14 prematurely arteriosclerotic patients who were considered to be heterozygotes for cystathionine synthase deficiency on the strength of pathological levels of homocysteine after the load, are compared with those data in 20 adult obligate heterozygotes and in 15 adult homozygotes. Table 4 compares the results of the determination of cystathionine synthase activity in skin fibroblast cultures in these 14 prematurely arteriosclerotic patients with the residual activities in the obligate heterozygotes and the homozygotes. Tables 3 and 4 illustrate that both the peak levels of homocysteine after methionine loading as well as the cystathionine synthase activities are of similar ranges in these 14 patients with occlusive arterial disease and the obligate heterozygotes. In addition, it is shown that in homozygous patients the peak levels of homocysteine after loading are manyfold higher whereas their cystathionine synthase activities are significantly lower ($P < 0.0001$). There is no statistically significant correlation between the individual residual cystathionine synthase activities and the after load peak levels of homocysteine in the group of 14 patients with premature arteriosclerosis ($r_s = +0.14$, $P^{++} > 0.10$), confirming the finding in obligate heterozygotes[6].

*Table 1.* Age, sex, peak levels of homocysteine after methionine loading, and angiographical and clinical features in 25 patients with premature occlusive peripheral arterial disease

| patients with occlusive peripheral arterial disease n=25 | | age yr | sex | peak homocysteine µmol/l | angiographically demonstrated arteriosclerotic lesions | presenting clinical signs |
|---|---|---|---|---|---|---|
| heterozygotes | no. 1 | 37 | ♀ prem. | 18.7 | aorta abdominalis | intermittent claudication |
| | no. 2 | 43 | ♀ prem. | 29.1 | aorta abdominalis<br>aa. iliacae | intermittent claudication<br>Raynaud symptoms |
| | no. 3 | 43 | ♀ prem. | 30.9 | a. carotis int.<br>a. anonyma<br>aorta abdominalis<br>a. renalis<br>aa. iliacae | renovascular hypertension |
| | no. 4 | 38 | ♀ prem. | 31.2 | aa. subclaviae,<br>a. anonyma<br>aorta abdominalis<br>a. renalis<br>aa. iliacae | vertigo<br>intermittent claudication<br>renovascular hypertension |
| | no. 5 | 49 | ♀ postm. | 39.4 | aorta abdominalis | intermittent claudication |
| | no. 6 | 48 | ♀ postm. | 50.7 | truncus coeliacus<br>aa. mesentericae | angina abdominalis |
| | no. 7 | 42 | ♀ prem. | 66.8 | a. renalis<br>a. mesenterica<br>aa. iliacae | renovascular hypertension |
| with normal handling of methionine | no. 8 | 38 | ♂ | 5.3 | a. renalis | renovascular hypertension |
| | no. 9 | 37 | ♂ | 7.6 | a. brachialis<br>a. iliaca | Raynaud symptoms<br>intermittent claudication |
| | no. 10 | 22 | ♂ | 7.8 | a. pulmonalis | lung embolia |
| | no. 11 | 37 | ♀ prem. | 8.3 | aorta abdominalis<br>aa. iliacae | intermittent claudication |

| | | | | | | |
|---|---|---|---|---|---|---|
| | no. 12 | 49 | ♂ | 8.4 | a. subclavia | vertigo |
| | | | | | a. vertebralis | Raynaud symptoms |
| | no. 13 | 43 | ♂ | 8.6 | aorta abdominalis | intermittent claudication |
| | | | | | aa. iliacae | |
| | no. 14 | 39 | ♀ prem. | 9.1 | a. poplitea | intermittent claudication |
| | no. 15 | 22 | ♀ prem. | 9.2 | aa. iliacae | intermittent claudication |
| | no. 16 | 27 | ♀ prem. | 9.3 | aorta abdominalis | renovascular hypertension |
| | | | | | a. renalis | |
| | no. 17 | 23 | ♀ prem. | 9.4 | a. subclavia | renovascular hypertension |
| | | | | | a. renalis | angina abdominalis |
| | | | | | truncus coeliacus | |
| | | | | | aa. mesentericae | |
| | no. 18 | 40 | ♀ prem. | 10.3 | aa. carotides int. | ischemic left arm pain |
| | | | | | a. vertebralis | |
| | | | | | a. subclavia | |
| | no. 19 | 49 | ♂ | 11.8 | aorta abdominalis | intermittent claudication |
| | no. 20 | 49 | ♂ | 12.0 | aorta abdominalis | intermittent claudication |
| | no. 21 | 23 | ♂ | 12.9 | a. renalis | renovascular hypertension |
| | no. 22 | 28 | ♀ prem. | 13.7 | a. renalis | renovascular hypertension |
| | no. 23 | 37 | ♂ | 13.9 | aortic valve | decompensatio cordis |
| | | | | | aorta thoracalis | intermittent claudication |
| | | | | | aorta abdominalis | |
| | no. 24 | 49 | ♂ | 14.6 | aa. iliacae | intermittent claudication |
| | no. 25 | 49 | ♂ | 18.8 | aorta abdominalis | intermittent claudication |
| | | | | | aa. iliacae | Raynaud symptoms |
| | | | | | a. femoralis | |
| control subjects | | 14 –42 | ♀ prem. | 5.0 – 9.1 (n=10) | | |
| n=40 | | 25± 3+ | | 6.8± 0.4+ | | |
| | | 45 –59 | ♀ postm. | 17.1 –34.9 (n=10) | | |
| | | 54± 1 | | 24.1± 1.9 | | |
| | | 22 –61 | ♂ | 7.8 –19.2 (n=20) | | |
| | | 42± 3 | | 13.3± 0.8 | | |

+ mean ±S.E.M.

*Table 2.* Age, sex, peak levels of homocysteine after methionine loading, and angiographical and clinical features in 25 patients with premature occlusive cerebrovascular disease

| patients with occlusive cerebrovascular disease n=25 | | age yr | sex | peak homocysteine µmol/l | angiographically demonstrated arteriosclerotic lesions | presenting clinical signs |
|---|---|---|---|---|---|---|
| heterozygotes | no. 1 | 28 | ♂ | 22.8 | a. cerebri media | hemiplegia, aphasia |
| | no. 2 | 45 | ♂ | 22.9 | a. cerebri media | hemiplegia, frontal syndrome, epilepsy |
| | no. 3 | 49 | ♂ | 23.6 | a. cerebri media | hemiplegia, aphasia, epilepsy |
| | no. 4 | 40 | ♂ | 25.7 | a. basilaris | Wallenberg syndrome |
| | no. 5 | 49 | ♀ postm. | 38.7 | a. cerebri media | hemiplegia, hemihypaesthesia |
| | no. 6 | 49 | ♀ postm. | 44.4 | a. cerebri media | transient ischemic attack |
| | no. 7 | 35 | ♀ prem. | 54.3 | a. carotis int.<br>a. cerebri anterior | transient ischemic attacks |
| with normal handling of methionine | no. 8 | 31 | ♀ prem. | 4.6 | a. carotis int. | transient ischemic attacks |
| | no. 9 | 40 | ♂ | 6.0 | a. carotis int. | transient ischemic attack |
| | no. 10 | 45 | ♀ prem. | 6.1 | a. basilaris | brain stem syndrome |
| | no. 11 | 46 | ♂ | 6.1 | a. basilaris | brain stem syndrome |
| | no. 12 | 36 | ♀ prem. | 6.8 | a. basilaris | brain stem syndrome |
| | no. 13 | 49 | ♀ prem. | 7.0 | a. carotis int.<br>a. vertebralis<br>a. basilaris | transient ischemic attacks |
| | no. 14 | 33 | ♀ prem. | 7.3 | a. cerebri media | hemiplegia, aphasia |
| | no. 15 | 40 | ♀ prem. | 7.3 | a. cerebri media | hemiplegia, aphasia |
| | no. 16 | 47 | ♂ | 7.6 | a. cerebri media | hemiplegia |
| | no. 17 | 44 | ♂ | 7.6 | a. cerebri posterior | hemianopsia |
| | no. 18 | 44 | ♀ prem. | 7.8 | a. carotis int. | transient ischemic attacks |

| | | | | | | |
|---|---|---|---|---|---|---|
| | no. 19 | 44 | ♂ | 8.0 | a. cerebri media | hemiplegia, aphasia |
| | no. 20 | 32 | ♀ prem. | 9.3 | a. carotis int.<br>a. cerebri media | hemiplegia, aphasia |
| | no. 21 | 42 | ♀ prem. | 10.1 | aa. carotides int. | transient ischemic attack |
| | no. 22 | 46 | ♂ | 12.5 | a. cerebri media | hemiplegia, aphasia |
| | no. 23 | 46 | ♀ postm. | 15.8 | a. carotis int. | transient ischemic attacks |
| | no. 24 | 34 | ♂ | 16.3 | a. cerebri media | aphasia |
| | no. 25 | 28 | ♂ | 17.4 | a. cerebri media | Gerstmann syndrome, aphasia |
| control subjects<br>n=40 | | 14 –42<br>25± 3⁺ | ♀ prem. | 5.0 – 9.1 (n=10)<br>6.8± 0.4⁺ | | |
| | | 45 –59<br>54± 1 | ♀ postm. | 17.1 –34.9 (n=10)<br>24.1± 1.9 | | |
| | | 22 –61<br>42± 3 | ♂ | 7.8 –19.2 (n=20)<br>13.3± 0.8 | | |

⁺mean ± S.E.M.

*Table 3.* Peak levels of homocysteine after methionine loading (ranges and means, ± S.E.M.) in 14 patients with premature occlusive arterial disease and pathological after load homocysteinemia, compared to those levels in 20 adult obligate heterozygotes and in 15 adult homozygotes for homocystinuria

| | Peak homocysteine after methionine loading µmol/l | | |
|---|---|---|---|
| | premenopausal women | postmenopausal women | men |
| patients with premature occlusive arterial disease and pathological after load homocysteinemia (n=14) | 18.7 - 66.8<br>38.5± 7.4<br>(n=6) | 38.7 - 50.7<br>43.3± 2.8<br>(n=4) | 22.8 - 25.7<br>23.8± 0.7<br>(n=4) |
| obligate heterozygotes (n=20) | 15.2 - 99.8<br>41.0± 7.1<br>(n=12) | 20.6 - 51.9<br>36.3± 15.7<br>(n=2) | 15.9 - 51.5<br>26.7± 5.2<br>(n=6) |
| homozygotes (n=15) | 131 -307<br>205.5± 25.3<br>(n=6) | 104 -290<br>221.6± 60.0<br>(n=3) | 92 -208<br>150.0± 19.9<br>(n=6) |

*Table 4.* Cystathionine synthase activities measured in fibroblast cultures in 14 premature occlusive arterial disease-patients with pathological homocysteinemia after methionine loading, compared to the activities in 20 adult obligate heterozygotes and in 15 adult homozygotes for homocystinuria

| | | Cystathionine synthase activity $nmol \cdot h^{-1} \cdot mg^{-1}$ | Peak homocysteine after methionine loading μmol/l |
|---|---|---|---|
| premature occlusive | | 1.9 | 66.8 |
| arterial disease - | | 0.4 | 54.3 |
| patients with pathological | | 4.5 | 50.7 |
| after load homocysteinemia | | 3.1 | 44.4 |
| (n=14) | | 4.1 | 39.4 |
| | | 0.3 | 38.7 |
| | | 1.2 | 31.2 |
| | | 0.4 | 30.9 |
| | | 2.4 | 29.1 |
| | | 2.7 | 25.7 |
| | | 0.7 | 23.6 |
| | | 0.6 | 22.9 |
| | | 1.4 | 22.8 |
| | | 2.2 | 18.7 |
| | range | 0.3 - 4.5 | |
| | mean | 1.9 ± 0.4 SEM | |
| obligate heterozygotes | range | 0.1 - 4.2 | |
| (n=20) | mean | 2.2 ± 0.2 | |
| homozygotes | range | 0.0 - 0.14 | |
| (n=15) | mean | 0.04± 0.01 | |
| control subjects | range | 3.6 -25.7 | |
| (n=23) | mean | 11.3 ± 1.1 | |

DISCUSSION

About 30% of a group of 50 unselected patients, under the age of 50, with clinically overt and angiographically proven occlusive peripheral or cerebral arterial disease were identified as heterozygotes for homocystinuria due to cystathionine synthase deficiency. In 25 patients with myocardial infarction under the age of 50 no heterozygotes were detected. The carrier state was assessed by a well standardized in vivo oral methionine loading test and in vitro measurement of cystathionine synthase activity in cultured fibroblasts from skin biopsies. The latter more complicated procedure was reserved for those patients in whom the former had shown pathological homocysteine accumulation. In an earlier study we provided ample evidence that pathological homocysteine accumulation after the methionine loading occurred in 90% of obligate heterozygotes, whereas the enzyme assay identified 85% of them[6]. It is therefore unlikely that by omitting the enzyme assay in the remaining patients heterozygotes might have been missed.

The probability to detect heterozygotes in the normal population has been estimated at 1 in 70 at the most[1]. We ourselves detected one heterozygote among 40 healthy controls. Therefore, the observed strikingly high frequency of heterozygotes in the groups of patients with premature occlusive peripheral arterial or cerebrovascular disease is unequivocal proof of increased risk of such pathology in carriers for homocystinuria. Additional known risk factors such as diabetes, hyperlipoproteinemia and essential hypertension were excluded in all patients studied, and, in the heterozygous and the non-heterozygous groups tobacco and oral contraceptives were used to a similar extent.

It is of interest that all carriers for homocystinuria in the group of patients with occlusive peripheral arterial disease under the age of 50 were women whereas the sex ratio in the non-heterozygous patients approximated the sex ratio for peripheral arterial disease in the Framingham study[11]. We do not have a straightforward explanation for this difference. In an earlier study on methionine handling in normal subjects we suggested that especially women in their reproductive life cycle might be protected better against vascular damage by their unique efficiency to metabolize methionine without accumulation of homocysteine[5]. We speculate that especially these women are liable to develop endothelial damage under the influence of homocysteine toxicity in the case of lower levels of cystathionine synthase.

The absence of heterozygotes in the myocardial infarction group is reminiscent of the report by Wilcken et al. who also failed to detect heterozygotes among patients with established premature coronary vascular disease[3]. In a recent multicenter study based on data of the morbid history of 629 patients with *homozygous* homocystinuria – the largest number of patients ever reported – Mudd et al. including two of us (G.B., B.F.), mentioned that 109 out of 253 registered thromboembolic events occurred in cerebral or peripheral arteries and only 10 produced myocardial infarctions[2]. From these data it is intelligible that *mild* homocysteinemia as present in *heterozygotes* does not lead to a significantly increased risk of heart attack in carriers and that we did not find any heterozygote in the group of patients with premature myocardial infarction. In this context it is intriguing that Mudd et al. in an earlier study based upon analysis of questionnaires from about 800 obligate heterozygotes estimated that less than 5% of homocystinuria heterozygotes might have a fatal or nonfatal heart attack by age of 50[12]. We ourselves in a relatively small group of 20 adult obligate heterozygotes established electrocardiographical signs of previous myocardial infarction in only one 43 year old carrier.

The remarkable difference in vulnerability between the endothelium of peripheral and cerebral arteries on one hand and the coronaries on the other remains as yet puzzling. In a recent study by de Groot et al. it was demonstrated that the enzyme cystathionine synthase is present in endothelial cells cultured from umbilical cord vessels from an obligate heterozygote for homocystinuria[13]. It seems possible therefore that endothelial cells, if deficient in this enzyme activity, may contribute to their own dysfunction by intracellular homocysteine accumulation. It seems of interest to explore the possibility that the endothelium from various areas of the circulatory system might show up differences in cystathionine synthase activity and consequently might differ in vulnerability by pathological homocysteine accumulation.

Treatment with pharmacological doses of vitamin $B_6$ as the cofactor for cystathionine synthase might prevent a number of serious clinical sequelae of homocystinuria. Very recently Mudd et al. elegantly showed that the number of thromboembolic events observed during pyridoxine treatment in biochemically vitamin $B_6$-responsive *homozygous* patients was far less than the number expected if such therapy failed to have an effect[2].
To our knowledge no study so far has shown that treatment with high-dose vitamin $B_6$ would stop vascular damage if present in *heterozygotes* for homocystinuria. Up to the present, in 10 of our 11 heterozygotes with premature vascular damage, homocysteinemia after methionine loading substantially decreased and even became normal in 6 of them after 6 weeks of treatment with vitamin $B_6$. The clinical relevance of this observation awaits long-term evaluation.

From our study it seems to be of practical interest to screen for heterozygosity for homocystinuria in patients with occlusive peripheral or cerebral artery disease under the age of 50 leading to common clinical syndromes such as intermittent claudication, renovascular hypertension and ischemic cerebrovascular pathology.

REFERENCES


1. Mudd SH, Levy HL. Disorders of transsulphuration. In: Stanbury JB, Wyngaarden JB, Fredrickson DS, eds. The metabolic basis of inherited disease. New York: McGraw-Hill, 1983, 522–59.
2. Mudd SH, Skovby F, Levy HL et al. The natural history of homocystinuria due to cystathionine β-synthase deficiency. Am J Hum Genet 1985; 37: 1–31.
3. Wilcken DEL, Reddy SG, Gupta VJ. Homocysteinemia, ischemic heart disease, and the carrier state for homocystinuria. Metab Clin Exp 1983; 32: 363–70.
4. Wilcken DEL, Wilcken B. The pathogenesis of coronary artery disease; a possible role for methionine metabolism. J Clin Invest 1976; 57: 1079–82.
5. Boers GHJ, Smals AG, Trijbels FJ, Leermakers AI, Kloppenborg PW. Unique efficiency of methionine metabolism in premenopausal women may protect against vascular disease in the reproductive years. J Clin Invest 1983; 1971–6.
6. Boers GHJ, Fowler B, Smals AGH et al. Improved identification of heterozygotes for homocystinuria due to cystathionine synthase deficiency by the combination of methionine loading and enzyme determination in cultured fibroblasts. Hum Genet 1985; 69: 164–9.
7. Boers GHJ, Smals AGH, Drayer JIM, Trijbels FJM, Leermakers AI, Kloppenborg PW. Pyridoxine treatment does not prevent homocystinemia after methionine loading in adult homocystinuria patients. Metab Clin Exp 1983; 32: 390–7.

8. Boers GHJ, Polder TW, Cruysberg JRM et al. Homocystinuria versus Marfan's syndrome: the therapeutic relevance of the differential diagnosis. Neth J Med 1984; 27: 206–12.
9. Schoonderwaldt HC, Boers GHJ, Cruysberg JRM, Schulte BPM, Slooff JL, Thijssen HOM. Neurologic manifestations of homocystinuria. Clin Neurol Neurosurg 1981; 83: 153–62.
10. Wicherink-Bol HF, Boers GHJ, Drayer JIM, Rosenbusch G. Angiographic findings in homocystinuria. Cardiovasc Intervent Radiol 1983; 6: 125–8.
11. Dawber ThR. Incidence of coronary heart disease, stroke, and peripheral arterial disease. In: Dawber ThR., ed. The Framingham study. Cambridge, Massachusetts and London, England: Harvard University Press, 1980, 59–75.
12. Mudd SH, Havlik R, Levy HL, McKusick VA, Feinleib M. A study of cardiovascular risk in heterozygotes for homocystinuria. Am J Hum Genet 1981; 33: 883–93.
13. De Groot PhG, Willems Ch, Boers GHJ, Gonsalves MD, van Aken WG, van Mourik JA. Endothelial cell dysfunction in homocystinuria. Eur J Clin Invest. 1983; 13: 405-10.

# Epilogue

## Part I HOMOZYGOUS HOMOCYSTINURIA

It has to be emphasized on the basis of data from the study on our group of homozygous homocystinuria patients, that the diagnosis should be considered not only in infants but also in adults presenting one or more symptoms suggestive of the homocystinuria syndrome. The majority of the patients shows striking clinical signs such as ectopia lentis, scoliosis, or marfanoid features, but the large heterogeneity in the expression of symptoms among homozygous patients should be kept in mind.
Also in the absence of the typical and well recognized homocystinuria symptoms, investigations for the presence of this disorder should be performed in patients suffering from, e.g. severe arteriosclerotic lesions and thromboembolic events at young age, or in children presenting psychomotor retardation or learning disabilities. Homozygous patients also may show striking psychiatric illness as the presenting symptom of homocystinuria.

A large variance also has been observed in abnormal biochemical characteristics among the studied homozygous patients. It has been concluded that most of them will show marked homocysteinemia and hypermethioninemia with lowered levels of cystine even in the fasting blood. Some exceptional homozygous patients, however, demonstrate levels of these amino acids in the fasting state very near to or even within the range measured in control subjects. Therefore, when patients present symptoms suggestive of homocystinuria, this condition should not be excluded unless a methionine loading test discloses a normal rise of homocysteine levels after the load. Final proof of homozygous cystathionine synthase deficiency is obtained by determination of the enzyme activity in cultured fibroblasts from a skin biopsy. Not detectable or only residual activities ranging from 1 to 3% of the mean control activity are in accordance with the homozygous state.

A word of grave warning is in place against the use of chemical diagnostic procedures other than quantitative amino acid analysis of blood. The use of semiquantitative or qualitative methods, e.g., the still popular urinary cyanide-nitroprusside test according to Brand and its modification as the silver-nitroprusside test, has to be considered unreliable because of their high score of false negative results, in our hands even in 40% of the homozygous patients.

As far as the treatment is concerned, it has been shown that high-dose pyridoxine treatment in so called completely pyridoxine responsive homocystinuria patients normalizes the abnormal amino acid profile in their fasting state, but does not prevent marked homocysteinemia after a major dietary methionine intake. Therefore, it is mandatory even in "complete" pyridoxine responders, to avoid major protein intakes from the diet and at least to distribute the dietary methionine more equally over the day. In "partially" or pyridoxine non-responsive patients additional treatment with folic acid and dietary methionine restriction should be prescribed, as these regimens induce a further decrease of homocysteine levels in such patients. Very recently, the use of betaine has been advised as a homocysteine-lowering treatment in pyridoxine non-responsive homocystinuria patients[1]. Its biochemical effect is based upon the enhancement of the re-methylation of homocysteine to methionine, leading to a decrease of the homocysteine levels and an increase of methionine concentrations. From our preliminary experience it is obvious that administration of betaine may induce a substantial lowering of the homocysteine levels not only in pyridoxine non-responsiveness as was reported by Wilcken et al[1], but also when pyridoxine treatment has led to an insufficient decrease of the homocysteinemia in so called partially pyridoxine responsive homocystinurics. A major increase of methionine levels as was warned against by Mudd et al[2], we observed in only one of the 4 patients we treated so far. Adverse effects of the long-term treatment with betaine have not been reported in literature up to the present[1]. The use of betaine could be preferable to dietary methionine restriction as an additional measure to reduce homocysteinemia.

The clinical outcome after the start with homocysteine-lowering treatment in our homozygous homocystinuria patients indicates a beneficial effect of this therapy in accordance with data provided by Mudd et al[2] in a recent collaborative study on more than 600 homozygous patients. However, in our group of patients the uneventful course may be partially due to the simultaneously prescribed treatment with anticoagulant drugs in 50% of them.

Part II HETEROZYGOUS HOMOCYSTINURIA

The excessive risk of premature vascular damage in heterozygosity for homocystinuria may have a sizable public health impact as the frequency of heterozygotes has been estimated so far as high as 1 in 70 to 1 in 200 in the general population. When future studies will have proven the efficacy of high dose pyridoxine treatment in patients suffering from arteriosclerotic lesions at young age due to heterozygosity for homocystinuria, its detection is not of merely academic interest. Therefore, efforts to develop a valid and simple test for the detection of the heterozygous state should be continued, because the methionine loading test - although a sensitive tool in our hands - is rather laborious and expensive to screen large numbers of subjects. A promising approach to a feasible test to identify heterozygosity can be the measurement of total homocysteine after disruption of all its disulphide bridges and its bindings to proteins after addition of dithiothreitol to a single blood sample. The total amount of measured homocysteine then is brought up by the concentrations of homocystine (homocysteine-homocysteine) and homocysteine-cysteine mixed disulphide as determined by conventional analysis, *plus* the protein-bound homocysteine. Pilot studies performed up to the present in our laboratory indicate that the in this way obtained total homocysteine level in a single fasting blood sample differentiates between homozygosity, heterozygosity, and the normal state[3]. If confirmed by future studies, the "total" homocysteine assay will provide a simple method for the screening of large numbers of subjects in order to directly establish the true incidence of heterozygosity in the general population, and, even more important so, to investigate on a routine base patients suffering from premature vascular damage or other signs suggestive of homocystinuria, for the presence of the homozygous or heterozygous state of this condition.

REFERENCES

1. Wilcken DEL, Wilcken B, Dudman NPB, Tyrrell PA. Homocystinuria. The effect of betaine in the treatment of patients not responsive to pyridoxine. N Engl J Med 1983; 309: 448–53.
2. Mudd SH, Skovby F, Levy HL, Pettigrew KD, Wilcken B, Pyeritz RE, Andria G, Boers GHJ, Bromberg IL, Cerone R, Fowler B, Gröbe H, Schmidt H, Schweitzer L. The natural history of homocystinuria due to cystathionine β-synthase deficiency. Am J Hum Genet 1985; 37: 1–31.
3. Boers GHJ, Trijbels JMF, Corstiaensen J, van Raay B, Smals AGH, Kloppenborg PWC. In preparation.

# SUMMARY

In part I of this thesis several aspects of homozygous homocystinuria due to cystathionine synthase deficiency are discussed. In part II attention has been devoted to the validity of methods for detection of heterozygotes for this condition as well as to the predisposition of heterozygotes to develop premature vascular pathology, the most serious complication in homozygotes.

In Chapter 1 the state of the art with regard to the biochemical, enzymatic, pathological and therapeutic aspects of homozygous homocystinuria has been surveyed on the basis of a number of references to the literature. The clinical effect of treatment for homocystinuria and the prognosis in the patients have been discussed with special reference to a recently published study in which the author of this thesis collaborated with Mudd et al., surveying the natural histories in more than 600 homozygous homocystinuria patients, including a subset of our patients (Am J Hum Genet 1985; 37: 1–31).

In Chapter 2 data on the morbid history are compiled with biochemical, enzymatic, pathological, and therapeutic features in 21 homozygous homocystinuria patients. It is pointed out in this Chapter that the studied group of patients is rather unique because of the fact that the diagnosis has been made in 15 of them after they had reached their adult age. All patients except one proved to be pyridoxine responsive. This high frequency of pyridoxine responsiveness in our group of homocystinurics contrasts with the reportedly rather high frequency of pyridoxine unresponsiveness in patients detected during infancy and childhood.

There is a large heterogeneity in expression of clinical signs. One of the patients does not show any clinical symptom up to the present at age of 32 years, whereas the single pyridoxine non-responsive patient proved to be also the most severely affected one, presenting almost all signs of the homozygous syndrome. Clinical symptoms such as scoliosis, ectopia lentis with myopia, and marfanoid features occurred as the most consistent findings in frequencies of 81%, 76% and 62% respectively. Moreover, further evaluation disclosed involvement of the vascular system in 86% of them, leading to major complications in 7 (33%), and osteoporosis in 14 patients (82%). Electroencephalographic abnormalities were found in 75%, albeit only minor and aspecific findings in most of the patients. Mild or gross mental retardation was a feature in a minority of the patients (24%). It is shown in this Chapter that after the start of therapy for homocystinuria the clinical course was uneventful up till the end of 1984 in all patients, except in 2, in one of which mental capacity deteriorated with increasing frequency of seizures and in another in whom retinal detachment occurred 6 years after the start of treatment.

Chapter 3 presents data on the results of a standardized oral methionine loading test in 8 of the 21 homozygous homocystinuria patients both without and during treatment with pyridoxine. Such therapy nearly normalized the pathological amino acid spectrum at least in the fasting state. However, after methionine loading the concentrations of homocystine in the serum of the patients while on treatment still rose substantially and the levels of homocysteine-cysteine mixed disulphide and cystine approximated those measured before the start with pyridoxine. This finding indicates that pyridoxine treatment attenuates the biochemical abnormalities in fasting pyridoxine responsive homocystinurics, but leaves their impaired capacity to handle major methionine loads essentially unchanged. In this Chapter it is stressed therefore that a major methionine intake with the diet should be avoided even in "completely" pyridoxine responsive homocystinuria patients.

In Chapter 4 it is pointed out that differentiating homocystinuria from Marfan's syndrome on the basis of the clinical symptoms may be very hazardous. Although there are a number of differences between both syndromes in their clinical expression, it is discussed that none of these is absolute. The wrong diagnosis of Marfan's syndrome in a homocystinuria patient would deprive him or her from appropriate treatment which has a favourable biochemical and clinical effect in the majority of homocystinurics. It is stated that quantitative serum amino acid analysis to exclude the possibility of homocystinuria should be performed in any patient suspected of Marfan's syndrome, especially when the autosomal dominant inheritance of the latter disorder is not apparent in the family. The relevance of making the correct diagnosis at young age has been illustrated in 2 homocystinuria patients, in whom the earlier incorrect diagnosis of Marfan's syndrome is held responsible for their dramatic clinical outcome.

In Chapter 5 it is emphasized that, based upon the rate of detection of homozygous homocystinurics in newborn screening programs, the frequency of heterozygosity for this condition is estimated to be as high as 1 in 70 to 1 in 200 in the general population. However, frequencies directly derived from groups screened for heterozygosity are not available as yet. Furthermore, in this Chapter data from literature on the pathogenetic effect of mild homocysteinemia as present in heterozygotes for homocystinuria, are discussed. It is shown that the conclusions from these data are strongly contradictory.

In Chapter 6 data are presented on the results of standardized oral methionine loading tests performed in normal subjects, divided in groups of pre- and postmenopausal women and groups of men of comparable ages. It is shown that there are statistically significant differences in methionine handling between pre- and postmenopausal women and men. Homocystine was detectable in the fasting serum in not any of the subjects. After methionine loading homocystinemia did not occur in any of the premenopausal women, in contrast to its marked presence in postmenopausal women and to a moderate extent in men. Moreover, in premenopausal women the levels of homocysteine-cysteine mixed disulphide were significantly lower than those in any other group, both in the fasting state as well as after methionine loading. It is concluded in that Chapter that premenopausal women have a unique efficiency of methionine handling, and thereby are preserved against accumulation of homocysteine after methionine loading. It is further speculated that this phenomenon might account for the lower incidence of vascular disease in women in the reproductive life cycle. It is also pointed out in this Chapter that a warning is in place to conclude to the presence of heterozygosity for homocystinuria in a postmenopausal woman merely on the strength of marked homocysteinemia after methionine loading.

In Chapter 7 the results of a standardized oral methionine loading test in 20 obligate heterozygotes for homocystinuria are shown. After load peak levels of total homocysteine – calculated as twice the concentration of homocystine plus the concentration of homocysteine-cysteine mixed disulphide – in the blood of 18 of the 20 obligate carriers exceeded those levels obtained in the appropriate control groups. Therefore, such after load peak levels discriminate 90% of the obligate heterozygotes from the normal controls. It has again been emphasized that this high score of identification of heterozygotes for homocystinuria only has been obtained by differentiating the obligate carriers and the control subjects according to sex and menopausal state. In this Chapter it is also pointed out that the more sophisticated procedure of measuring cystathionine synthase activity in cultured fibroblasts from these 20 obligate heterozygotes identified 17 of them by the presence of lowered enzyme activities with a mean level of 19% of the mean control activity. Taking the results of both tests together all 20 obligate heterozygotes could be decisively identified.

In Chapter 8 the outcome of screening for heterozygosity in a group of patients with premature arteriosclerosis has been presented. Seventy-five patients, all under the age of 50, have been investigated, 25 suffering from occlusive peripheral arterial disease, 25 with occlusive cerebrovascular disease, and 25 with myocardial infarction. In these patients known risk factors of arteriosclerosis such as diabetes, hyperlipoproteinemia, and essential hypertension had been excluded. Heterozygosity for homocystinuria could be established in about 30% of the patients in each of the former groups, whereas in the group of 25 patients with myocardial infarction not any proved to be heterozygote. Because of the frequency of heterozygosity for homocystinuria in the normal population of 1 in 70 at most, it is concluded that heterozygosity predisposes to the development of premature occlusive arterial disease, leading to common clinical disorders such as intermittent claudication, renovascular hypertension, and ischemic cerebrovascular pathology.

Bei Fragen zur Produktsicherheit wenden Sie sich bitte an:
If you have any questions regarding product safety,
please contact:

Walter de Gruyter GmbH
Genthiner Straße 13
10785 Berlin
productsafety@degruyterbrill.com